SEAGRAVE

A Pictorial History of Seagrave Fire Apparatus

by Matthew Lee

Additional copies of this publication available from:
Matthew Lee, P. O. Box 55, Plymouth, Michigan 48170

ISBN 0-9630472-0-5

Printed by Industrial Graphics Service, Kalamazoo, Michigan
Electronic Publishing by The Fix Edge, Belleville, Michigan

A collection of name plates from Seagrave's past

In the first column: a hand-drawn, a horse-drawn and a teen's motor version.

In the second column: an oval version from the 1920's through the early 1950's and its replacements from the 1970's and 1990's.

The FWD Seagrave Company provides photos and background information, on older apparatus, where possible. They have many, but not all, records dating back over the past 100 plus years. They can help you gather technical information, but locating and researching specific records takes time. For information on a specific piece of apparatus, write the FWD/Seagrave Museum, 105 East 12th Street, Clintonville, Wisconsin, 54929.

About the Book

This pictorial history book is an accumulation of facts and photos from many sources. I have been collecting photos of fire apparatus, for years, as a sideline to restoring apparatus. A few years back, I accumulated enough photos of Detroit apparatus to publish a pictorial history of the Detroit Fire Department. Creating the book was enjoyable and it was well received. The local Seagrave distributor asked if I'd be interested in publishing a Seagrave history book. This book is the result of that inquiry.

A few trips to Clintonville, Wisconsin, and seven trips to Columbus, Ohio, were just the start. Letters by the score and hundreds of phone calls netted about 2,000 photos and many pounds of literature. Over about two years, these items were assembled, along with commentary from, present and former, employees and experts in the fire apparatus hobby, as listed below.

Former employees of Seagrave

Andy Ayers	Sales Staff - 10 years of service.
Willard Brenning	Specification writer - 35 years of service
Lillian Fornoff	Presidential secretary - 35 years of service
Bob Moore	Seagrave repairman - 35 years of service
Paul Reinert	Seagrave repairman - 30 years of service
Bob Reith	Pump builder - 35 years of service
Sherlie Stone	Former sales manager - 30 years of service

Hobbyists who provided background information

Marvin Cohen	Secretary SPAAMFAA
John Gambs	Apparatus restorer, historian
Bill Hall	Historian, museum curator
Dan Martin	Photographer, historian, author
Walter McCall	Historian, author, lecturer
Walt Pittman	Historian, photographer, painter
Ken Soderbeck	Restorer, historian, painter
Jerry Strong	Apparatus delivery engineer
Jane Seagrave Sturdevant	Seagrave family historian

Historians and Photo Collectors

Richard Baker	Frank Heinlein	Sally Sieslik
Mark Berg	John Holden	Steve Sokol
Roland Boulet	Tom Kniaz	Wayne Sorensen
Al Burch	Jim Leininger	James Soteriades
Joe Burch	Jack Lerch	Norman Spain
Chuck Buschardt	Ed Levy	Eric Sprenger
Gregg Cariddi	Dale Magee	Richard Story
Bill Celentano	Ted Main	John Sytsma
Jack Conners	Frank Morgan	Knute Thompson
Bill Craven	Bill Noonan	Clarence Woodard
Ted Elder	John Peckham	Chuck Yoeman
Steve Hagy	Marvin Scholl	Murray Young

The Seagrave Company

The fine people of Seagrave have been very supportive in this effort. You could sense the feeling of pride in the office and plant personnel as I quizzed them on almost forgotten details. They spoke of these pages being not only the history of Seagrave, but the evolution of fire apparatus in America.

Additional historical information was provided by the Motor Vehicle Manufacturers Association and the Ford Motor Company.

Also, a special thanks to my sons, Ed and Jim, and my wife, Nancy, who provided input, advice and hours of help.

Many books have captions explaining each photograph in great detail. I chose not to use standard captions, but to discuss the significance of a photo or a set of photos. There are over 800 pictures or sketches in this book. To explain pump size or ladder length of each apparatus would have been redundant.

A text loaded with historical facts and background information, on the other hand, creates a setting that actually highlights a photograph. With proper historical perspective the reader can then enjoy and explore many aspects of each photo. The text in this book is meant to provide information that enhances, not just explains the photographs.

OFFICE OF

Seagrave & Co.,

MANUFACTURERS OF

FIREMEN'S GOODS,

Fire Escapes, Trussed Ladders, Trucks, Hose Wagons, &c.

416-418 MICHIGAN AVENUE,

TELEPHONE 951-3 RINGS.

C E Bunder Esq C.F.D. Detroit, Mich., Jan 18 189

H. B. SPAIN, ASS'T TO PRESIDENT — C. S. GREENLEE, PURCHASING AGENT — WM. E. LEE, GEN'L SUPERINTENDENT

NOT IN ANY TRUST OR COMBINATION

The Seagrave Company,

MANUFACTURERS OF STRICTLY HIGH GRADE

MOTOR PROPELLED FIRE DEPARTMENT APPARATUS OF ALL KINDS

ORIGINATORS OF CENTRIFUGAL FIRE PUMPING ENGINES

OFFICE AND FACTORY COLUMBUS, OHIO

THE SEAGRAVE COMPANY

ORIGINATORS OF CENTRIFUGAL FIRE PUMPING ENGINES

JULIUS F. STONE, PRESIDENT
H. B. SPAIN, VICE-PREST & GEN'L MGR.
CATHARINE COHEN, SEC'Y & TREAS.
C. S. GREENLEE, PURCHASING AGENT
WM. E. LEE, GEN'L SUPERINTENDENT

MOTOR PROPELLED FIRE APPARATUS

OFFICE AND FACTORY

COLUMBUS, OHIO

Address all Communications to the Company

THE SEAGRAVE CORPORATION

ORIGINATORS OF CENTRIFUGAL FIRE PUMPING ENGINES

MOTOR PROPELLED FIRE APPARATUS

OFFICE AND FACTORY

COLUMBUS. OHIO

The Seagrave letterheads reflect the companies history

Over 100 years ago, Frederic Seagrave started making ladders in a shed in rural Rochester, Michigan. His first customers were local apple orchard owners. The product was well received and prompted a move of forty miles to Detroit to be closer to supplies and markets. Here Frederic Seagrave, in 1881, set up shop and created Seagrave & Co. for the express purpose of selling fireman's goods, mainly ladders and wagons.

Detroit, between 1870 and 1890, saw a change from a lumber processing town to a steel town. City directories listed forty-two new manufacturing concerns that evolved in this twenty year period. The "lumber money", that fueled this boom, was now being invested in the manufacture of stoves, ships and rail cars. Frederic Seagrave knew, to gain more venture capital and space for his growing business, he would have to move.

In the buggy manufacturing town of Columbus, Ohio, Mr. Seagrave found an interest in his product, manufacturing space and working capital. The move to Columbus, Ohio was started, in 1891. By 1898, Mr. Seagrave's products attracted the eye of Julius Stone, a local banker, Ohio State Regent, and owner of The Ohio Buggy Company. Together they formed The Seagrave Company. That partnership lasted until 1902 when Mr. Stone bought out Frederic Seagrave and moved the business into the buggy factory at 2000 High Street, Columbus, Ohio. This was to be the home of the company for the next sixty-three years.

The Seagrave Company would remain closely held by Julius Stone and some local investors until 1925 when capital, again, was needed and a group of investment bankers bought into the company. The name was then changed to The Seagrave Corporation.

Stone remained active in the company, but daily operations were now handled by Howard Spain, General Manager. Seagrave weathered the depression, though sales and employment dropped from 300, in 1929, to 145, in 1935. A post depression upturn lasted through the war years.

Seagrave ended the year 1949 with backorders for equipment totalling over two million dollars. With record profits, the corporation redesigned equipment and looked for further expansion. Seagrave purchased a supplier, Fyr-Fyter, in 1952. They also purchased a competitor, Maxim Motor Company, in 1955. Further diversification came, in 1957, with the purchase of a modular home construction company and succession of other diverse manufacturers. Paint, building materials, and leather processing manufacturers became partners with fire apparatus in The Seagrave Corporation.

By, 1963, the fire apparatus side of the business needed a capital outlay for a new plant and tooling. It was profitable, but not as profitable as some of the nine other divisions. So, in mid 1963, the fire engine operation in Columbus was sold to The FWD Corporation.

Sadly, the very division that created the profit to purchase the other product divisions was sold.

FWD, a long time maker of specialty vehicles, as well as fire engines, struggled with the ancient plant and facilities, and reluctantly moved the operation to its headquarters in Clintonville, Wisconsin, in 1965.

Today, fire apparatus is FWD Seagrave's main business. The name Seagrave has appeared on fire apparatus for over 110 years.

And whatever became of the old Seagrave Corporation? A name change, to Vista Industries, was made to separate the new company from the association with the fire equipment business.

THE SEAGRAVE CORPORATION

122 EAST 42ND STREET, NEW YORK 17, N. Y.

OXFORD 7-7672

OFFICES AND PLANTS
COLUMBUS, OHIO
SUPRADUR MFG. CORP.
WIND GAP, PA.
MAXIM MOTORS CO.
MIDDLEBORO, MASS.

Seagrave

FIRE APPARATUS INCORPORATED

COLUMBUS, OHIO 43207

Seagrave

FIRE APPARATUS DIVISION FWD CORPORATION

2000 SOUTH HIGH STREET
COLUMBUS, OHIO 43207

---

seagrave

SEAGRAVE FIRE APPARATUS, INC.

CLINTONVILLE, WISCONSIN 54929

Phone: (715) 823-3194
Telex: 26-3424

Welcome to the World of Seagrave

Clintonville, Wisconsin

Nestled in the rolling land of Northeast Wisconsin is the industrious community of Clintonville. Settled by immigrants from Europe, the Community remained agriculturally oriented, until the advent of FWD, in 1908. Now both industry and agriculture play important roles.

The FWD Seagrave plant or "The Drive", as the locals refer to it, encompasses eleven acres in the heart of town and many people walk to work. Clintonville's history includes not only FWD Seagrave, but Northwest Airlines, as the local airport was home to Wisconsin Central Airlines . . . the humble beginnings of today's Northwest Airlines.

A word about FWD

Although, this book is about Seagrave, the names FWD and Seagrave today are synonymous, therefore, a word about FWD is appropriate. In 1907, Otto Zachow and his brother-in-law, William Besserdich, blacksmiths in Clintonville, designed a ball and socket that made it possible to steer powered wheels.

They reworked a Reo touring car into a steam powered four wheel drive, go anywhere vehicle. The adventure with steam was short lived, but the four wheel drive mechanism became the basis for The Four Wheel Drive Auto Company.

The U. S. government saw potential and encouraged development of the system. The federal government, while supportive, was slow to give up its mules. So the truck that was designed for the Army was sold to the Silver City Beer and Ice Company of Denver, Colorado, in 1913.

Company growth was steady and product lines expanded. In 1959, the name of the company was changed from The Four Wheel Drive Auto Company to the FWD Corporation. Further expansion was to include the Seagrave Fire Apparatus division of The Seagrave Corporation, in 1963.

Table of Contents

Seagrave has built thousands of pieces of apparatus and rebuilt hundreds more. It's hard to categorize production that included such a diverse enterprise. For example, Seagrave was listed in a local wagon builder guide as late as 1916, and often rebuilt twenty year old pieces of apparatus. The chapters outlined above are an attempt to list products of like appearance. Pumps, motors and other hardware would likely not follow this guide. Dates listed for apparatus in this book, for the most part, are dates gleaned from painter's logs or delivery lists.

The Early Years

1880 - 1912

Frederic S. Seagrave
1849 - 1923

In 1729, John Seagrave and his wife Sarah, left England for the new world. John died during the crossing, but Sarah arrived safely with two sons and two daughters. Five generations later, in Needham, Massachusetts on June 16, 1849, Frederic Scott Seagrave was born. He was the seventh of ten children. Disliking farming, but enjoying mechanics and adventure, Frederic moved West. At age 20, he married Adelaide Ruttledge of Detroit, Michigan. They had three sons. The middle son, Warren Edmund, would eventually be involved in the fire apparatus business. In 1874, the city directory lists Frederic as a commission merchant, or salesman. The 1880 census shows Frederic as living in nearby Rochester, Michigan with his family; mother-in-law, brother-in-law, and two sisters-in-law.

1881 marked the issue of his first patent for a trussed ladder. This first ladder had plain straight sides and was trussed with wires corner to diagonal corner. From 1882 through the next twenty years, a dozen patents for ladders, fire equipment, and mechanical devices would be credited to Seagrave. Homer Elliot and Seagrave, even worked together on what was to be the Elliot style truck axle, still in use today.

Seagrave & Co. grew along with Detroit in the 1880's. The apparatus gained in size and recognition. Soon, the business found itself competing for space and venture capital in Detroit. Frederic Seagrave found both, as well as willing wagon craftsmen and more favorable weather in Columbus, Ohio. The move took place over two years, 1891 to 1893. Frederic Seagrave was now the manager and tinkerer, and his son, Warren Edmund was the traveling salesman. As the quality of the apparatus, and the demand for it continued to grow, the little plant on Lane Avenue, at the Columbus Hocking Valley & Toledo Railroad track was too small.

Frederic, now middle aged, faced the decision to expand again, or sell and continue to invent and invest. A willing buyer was found in a local entrepreneur. Julius Stone, a Michigan farm boy, former railroad mechanic, and now successful Columbus businessman, bought into Seagrave & Co. in 1898. That year, the name was changed to The Seagrave Company. Frederic Seagrave retained the right to make and sell fire apparatus in Canada. In 1901, Frederic Seagrave set up a company named W. E. Seagrave (Warren Edmund) in Walkerville, Ontario, across from the river from Detroit to act as the Canadian Seagrave agent.

City directories list Frederic Seagrave as a patentier and investor until 1923. Frederic Scott Seagrave died on February 7, 1923, and was buried in the family mausoleum at the Windsor Grove Cemetery, Windsor, Ontario.

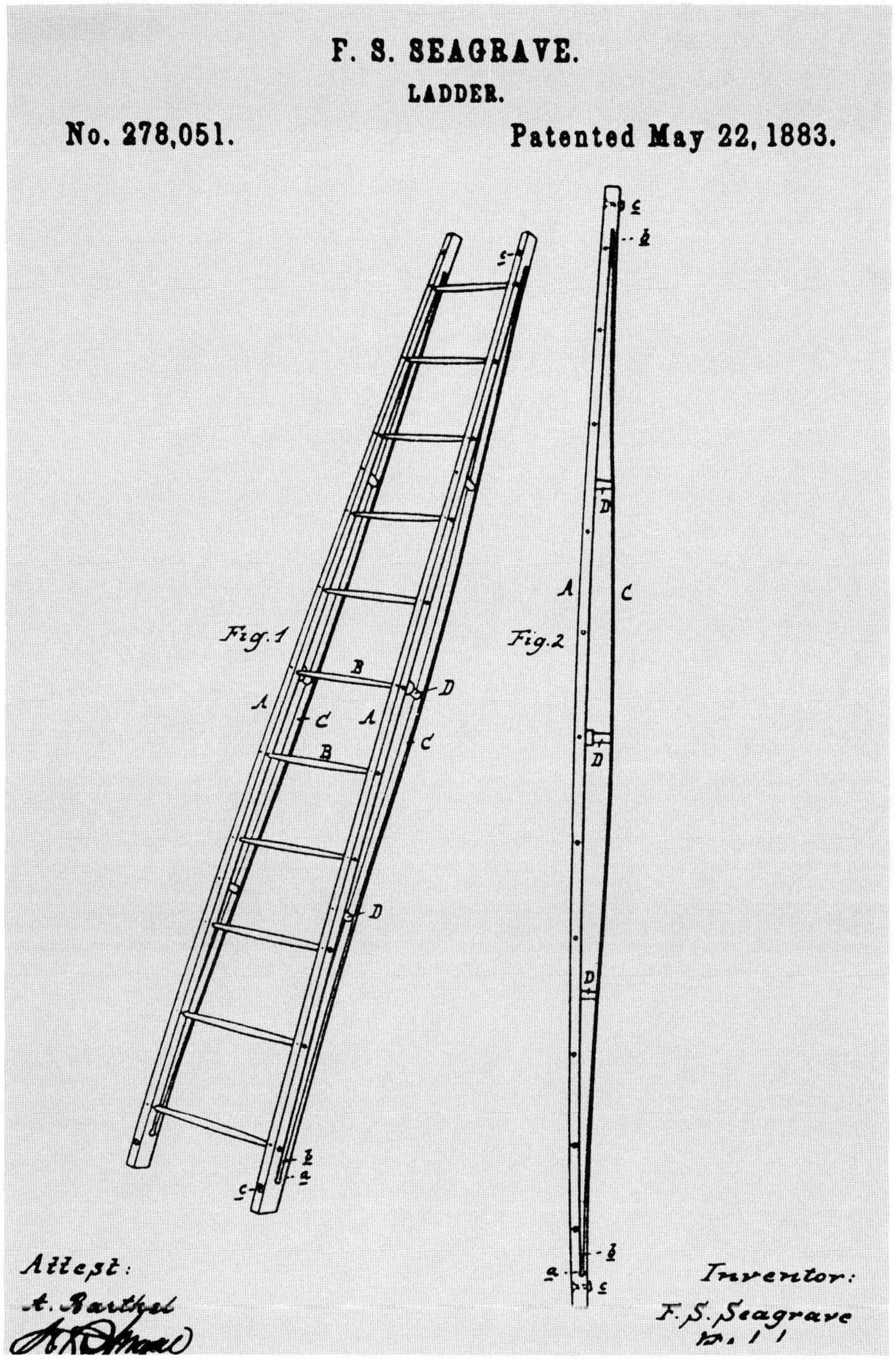

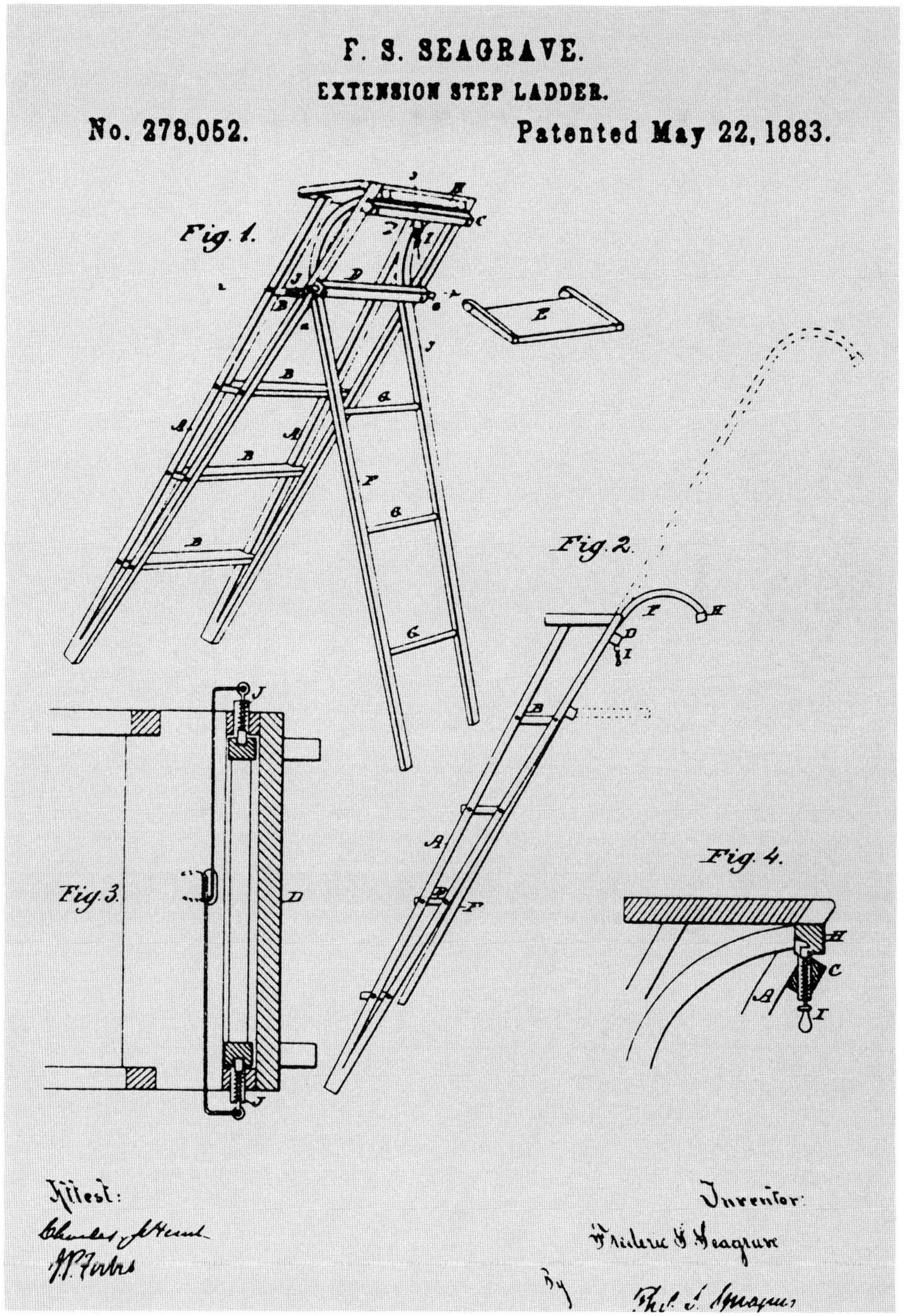

The first Seagrave patent issued in 1881 showed a ladder, reinforced corner to corner with wires. By 1883, the famous truss ladder was created. Seagrave style truss ladders are still produced today.

Trussed Firemen's Wall or Single Ladders.

6 to 18 feet................per foot,	$0 60	36 to 40 feet................per foot,	$1 00
20 " 26 " .... "	65	42 " 45 " .... "	1 25
28 " 30 " .... "	75	46 " 50 " .... "	1 50
32 " 34 " .... "	80	.. ..	....

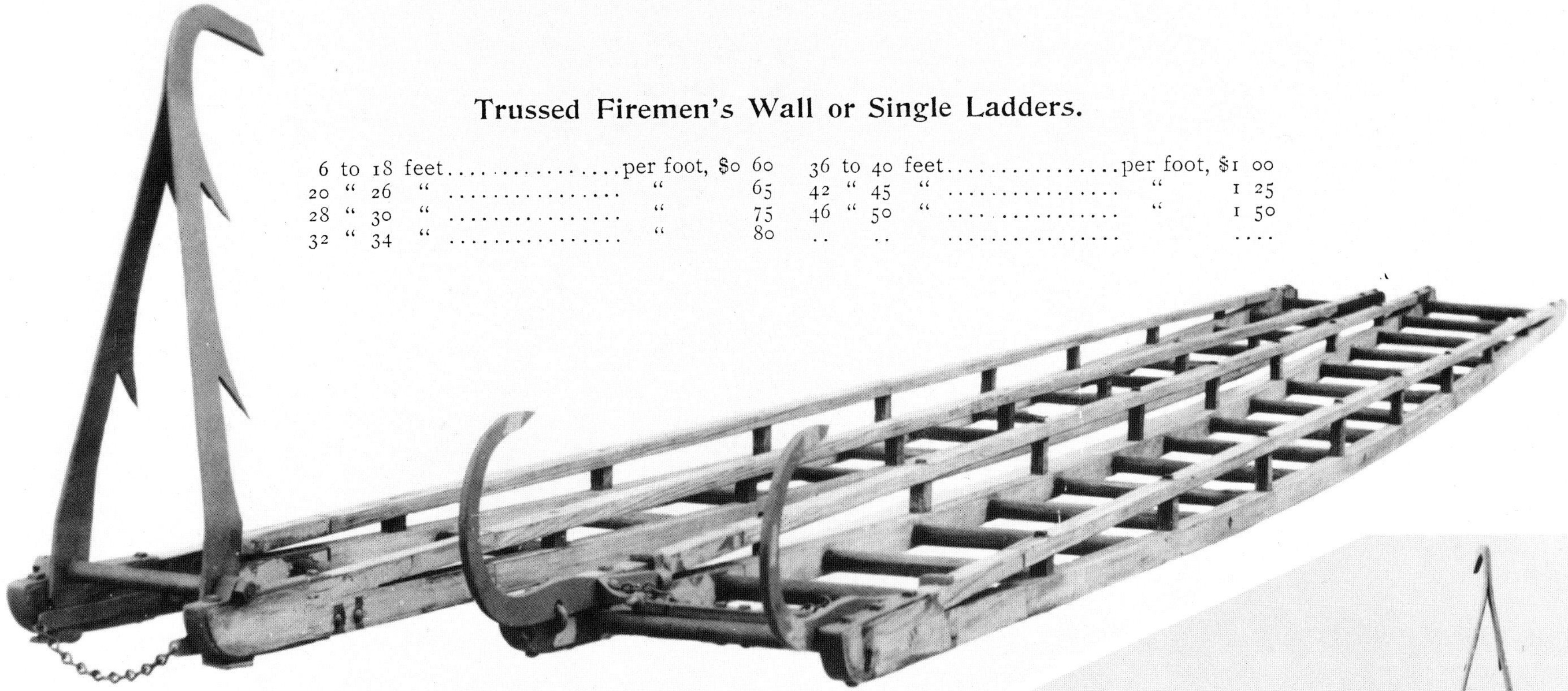

Firemen's Inside Extension Ladders..................per foot,		$1 25
Improved Scaling Ladders............................	"	1 50
Roof Ladders, with Special Hook....................	"	1 00
" **Regular Pattern**	"	75

Seagrave's Trussed Windlass Extension Fire Ladders.

With best pliable wire cable and improved union hooks, worked at pleasure from the ground and with stay poles. The Ladder is about one-half the weight and three times stronger than old style ladders.

40 ft., extends to 35 ft.....	$ 75 00	66 ft., extends to 60 ft.....	$175 00	86 ft., extends to 80 ft ...	$300 00
45 " " 40 ".....	90 00	70 " " 64 ".....	200 00	90 " " 83 ".....	350 00
50 " " 45 ".....	110 00	76 " " 70 ".....	225 00	97 " " 90 ".....	400 00
56 " " 50 ".....	130 00	80 " " 74 ".....	250 00	100 " " 94 ".....	450 00
60 " " 54 ".....	150 00	.. ..		... .	

All Firemen's Ladders ironed and painted vermilion or blue, and varnished.

Our New Rapid Hoist Appliance, used in place of Windlass, at same price, on all Ladders, unless otherwise ordered.

This page is a collage of information from an 1884 catalog. Close examination of the ladders show many features still in use today.

The photo at the right shows the 66 foot extension ladder displayed on a neighbors building. The ease of handling a scaling ladder is shown at the far left. This is the type of shed Seagrave used to store apparatus while the painting progressed.

OFFICE ON
201-225
.H. HABERKORN & CO.

ADDRESS CORRESPONDENCE TO SEAGRAVE & CO. DETROIT, MICH.

The basic village ladder trucks (left), carried roof and wall ladders, fire buckets, and pull down hooks. This was a model 5, and it cost $400. Equipment consisted of: a 40 foot extension ladder or 30 foot straight ladder, an assortment of ground ladders, and rubber fire buckets. Seagrave apparatus always featured lights at the head of the unit, or "headlights".

The improved model 7, shown here, came with the normal complement of a model 5 plus a pike pole, axe, torch, and warning gong. It weighed 700 lbs. and cost $600.

Ladders carried on a model 7 were:

- 40 foot extension
- 14 foot roof
- 16 foot roof
- 12 foot scaling
- 18 foot wall

Plus:

- 2 headlight lanterns
- 2 fire axes
- 2 pike and ladder lifting poles
- fire buckets
- pull down hook
- rope and chains

The demonstrator above is a model 7A. It was equipped like a model 7, but with the addition of soda acid fire extinguishers. Cost $700.

The horse-drawn truck is a model 8. They are furnished with horse shafts or hand poles. Ladders are: 50 ft. extension, 28 ft. wall, 20 ft. wall, 16 ft. roof, 12 ft. roof, and a 16 ft. scaling ladder. Cost $1,000.

The plain, unsprung ladder trucks on page 12 were referred to as village ladder trucks. As the weight and size of the trucks grew, the apparatus became city service ladder trucks.

The small wheels in front of the rig allowed the front axle to pivot more. The front wheels now would fit under the vehicle frame. The vehicle above is a model 8. They featured 2 inch wide wheels, but no braking system. The truck for Barnesville is a model 9. It has brakes on the rear axle.

These "city service" ladder trucks weighed 1,700 to 2,000 pounds. Brass trim was a $100 option, as was a rear steer tiller.

The building shown is on Elizabeth Street, about a block from the main office. Buildings were rented all over town to store partially painted apparatus as the varnish of the day could take a month to dry.

FROM
SEAGRAVE & Co.

SOMERVILLE F.D.
FROM
SEAGRAVE & Co.

The photos on these pages are model 9A's. They are model 9's with the inclusion of fenders and a wire basket beneath the belly of the apparatus, and running boards.

Typical complement of ladders were: 60 ft. extension, 40 ft. extension, 28 ft. wall, 20 ft. wall, 16 ft. roof and a 12 ft. roof. Additional equipment included: an assortment of pitch forks, wire cutters, pike and lifting poles, crow bars, buckets, axes, fire extinguishers, extra charges and hooks. Shown, above, are the hooks that give a "hook and ladder" its name.

The curved hook was a pull down hook used to remove porches, sheds or any structure that might provide a fire break. A pike pole was inserted into the hollow tubular base of the hook so that it could be placed, from a distance, onto a burning structure. Ten feet of chain led to 100 feet of stout rope so that many men or horses could be enlisted into the removal process.

The angular hook was used to hoist hose or equipment to upper floors. A pulley in its base aided in the task.

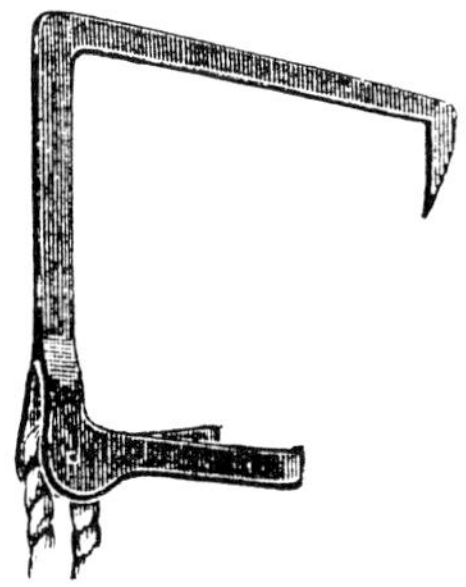

Optional equipment

A Detroit door opener like the one held by the painter in the photo above.

Improved Hose-Hoisting or Life-Saving Hooks.

The Improved Hose-Hoisting or Life-Saving Hook is similar to a Scaling Ladder Hook, and is thrown into a window or onto a cornice, and with a long rope running through a pulley at the bottom of the Hook, hose may be hoisted or persons rescued or let down by firemen to the ground.

Improved Hose-Hoisting and Life-Saving Hook, plain, with 150-foot Rope...........$15 00
" " " " nickel-plated, with 150-foot Rope.... 20 00

Frederic Seagrave owned Seagrave & Co. and the New England Paint and Color Company, next door. The address for Seagrave in Detroit was 418 Michigan Avenue, just down the street from what is now Tiger Stadium. That's Frederic in the doorway, showing off a working model of an aerial ladder. Notice the standard truss ladder nested in the frame of a larger trussed aerial ladder frame.

Sarah Stewart, a sister-in-law, managed the paint store. She would notify Frederic when customers arrived to look over fire equipment. Mr. Seagrave was always in between the workshop on Orchard Street and the many rented out buildings used as drying sheds.

Notice the only paving on Michigan Avenue, at the time, was between the streetcar tracks. Cobblestones kept the horses, pulling the streetcars, from gouging the dirt between the rails.

The move to Lane Avenue building in Columbus, Ohio, took place in 1891. They moved everything, including the sign from in front of the store. Note the same sign in both photos. By moving to Columbus, Ohio, the company gained in four areas. First, more space. The boom town of Detroit had doubled in size during the previous 15 years. Space for storing wagons while the paint dried was at a premium. It was not uncommon for a paint job to take a month or more. Second, more favorable weather. An extra two or three weeks drying time on each end of the winter season. Three, close to wagon parts. Columbus was the buggy capital of the Midwest. Two large local manufacturers were turning out a completed wagon every five minutes. And lastly, venture capital. Money in Detroit was going for iron products: stoves, ships, rail cars and steam engines. Columbus investors, on the other hand, were eager to invest, in a wagon business.

Seagrave's main office and research department remained at this location until 1910. That year a new three story building was built at the High Street location to house the expanding business. The Lane Avenue location is now near the heart of the Ohio State University campus.

The new building on Lane Avenue provided space and a siding on the Hocking Valley railroad. Rail was the prime delivery method. Close study of the main building and out-building will give the historian clues to fix other photos in time. Note, the photos on the next few pages will echo the buildings and the style of sign board used during this period.

Standard colors at this time were: white with black or gold trim, red with gold trim, or green with gold trim. Paint color and trim characteristics of each delivery was kept on file at the plant. Subsequent orders from the communities would usually contain a note . . . "paint just like last years delivery."

Perry, Iowa, received the white ladder truck, at left, complete with lanterns, an option at three dollars each. The torches on the above delivery were also a three dollar option. Other prices of the day were: 2 1/2 inch fire hose double jacket - 85 cents a foot, single jacket - 75 cents a foot, 36 inch play pipe - fifteen dollars, brass hook and spanner wrenches - eighteen dollars a dozen.

The model 9A, shown on this page, cost $1,500 plain and $1,600 with tiller rear steer. The 1/2 x 2 inch steel rims and 2 inch axles were standard. With a complement of equipment, including two fire extinguishers and twelve extra charges, these rigs weighed 2,500 pounds.

Foot operated gongs on these rigs later gave way to locomotive style bells as the gong was easily confused with the gong of the street car.

The gongs were made by a company called New Departure. Most of the company's first products were new departures from earlier methods, hence the name. New Departure is probably best remembered for ball bearings and the bicycle coaster brake.

Western states referred to these rigs as truck companies. This 1898 delivery to Los Angeles, California was truck 'C', shop number 75. It ran with Engine 11. In 1922, this veteran was motorized with a used Christie tractor, and served a few more years.

Some departments never know what size horses would be available, so they ordered one and two horse hitches. The typical dray or large draft horse was not fast enough for fire service. Racing type horses had the speed, but lacked the stamina. The mid sized Percheron breed was a favorite fire horse. Their color ran from a black dapple to white. Often, the grey spots had black centers.

Hose and ladder wagons were favorites of many small towns. The bed under the ladder carried 400 feet of 2 1/2 inch hose.

Charlotte, Michigan ordered this city combination wagon shown at the left, with room for 800 feet of 2 1/2 inch hose. This classic design: gong up front, ladders and extinguishers on the side, and a tool box under the rear step, became the standard for the next century.

Hose wagons with hose capacity of 800 feet or less could be hand or horse-drawn. The change meant removing the horse hitch and seat in favor of the drag rope. Because of the weight involved with a fully loaded wagon, these "convertible" models came with a brake pedal mounted on the rear step.

Seagrave wagons carried a five year guarantee. The majority of components were made in house, but specialized items were purchased. Seagrave used Meyer ball bearing axles and Sweede's steel springs.

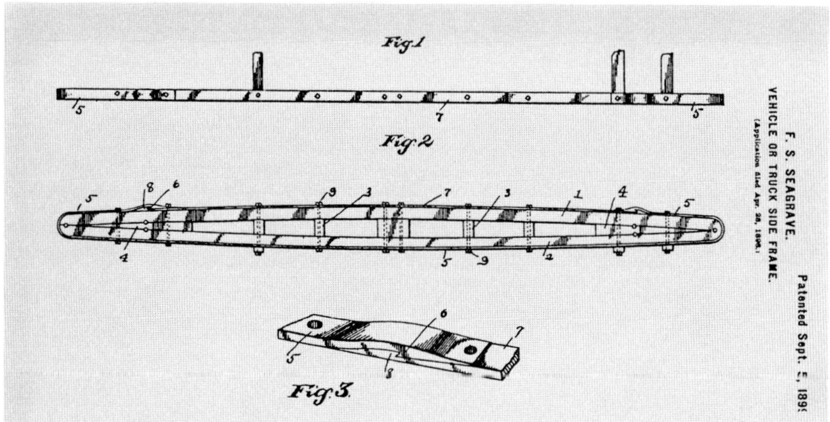

A steel band around the trussed wood frame was a Seagrave trademark. The band was heated to expand it, then it was placed on top of the wood frame. Cooling and contracting of the steel band created a rigid, shake-proof frame. The patent shows the installation of the steel band around the frame. These, before and after, photos show how all items were fitted, then disassembled for painting.

Straight hose wagons, hose and ladder wagons, and triple combination wagons (chemical, hose and ladder) showed the diversity of this custom builder. The typical wood body with a ventilated hardwood floor carried 800 feet of 12 1/2 inch hose. The standard hose wagon sizes carried 600', 700', 800', 1,000', or 1,250' of 2 1/2 inch fire hose.

The spring loaded horse hitch or wiffletree was another Seagrave patented invention. The truss and spring mechanism reduced road shock that normally would be transferred to the horse.

Rim size ran according to weight. As these ladder trucks grew in size, 2 1/2 inch and then 3 inch width rims became common. Rubber tired fire equipment would not appear for a few more years (see page 42). The rubber, being soft, allowed small stones to embed themselves in the rubber as the rig ran over them. As opposed to the steel rim that had to ride over even the smallest rock, the resilient, embeddable feature gave a smoother ride.

The Ohio State house, provided a backdrop for this delivery photo. It also appeared in ads for the four other wagon companies in Columbus at the time. They were:

The Columbus Buggy Co.
Ohio Carriage Manufacturing Co.
John Immel & Sons
United State Carriage Co.

Another imposing structure in Columbus is engine house No. 16. The 1904 structure is being restored. It houses the Central Ohio Fire Museum. Displays at the museum feature the Seagrave product line. There are no standard museum hours. For information, call the museum office at (614) 464-4099.

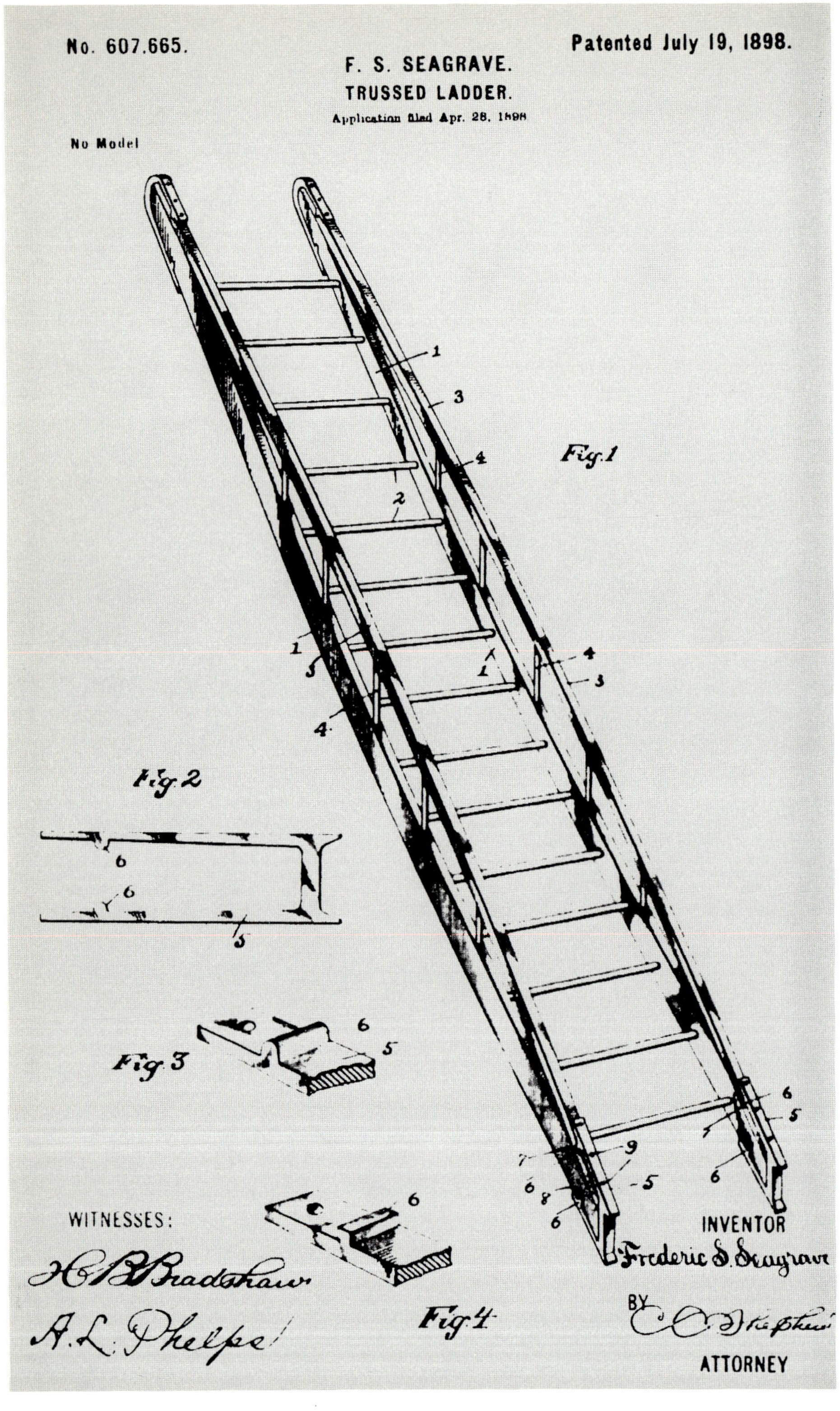

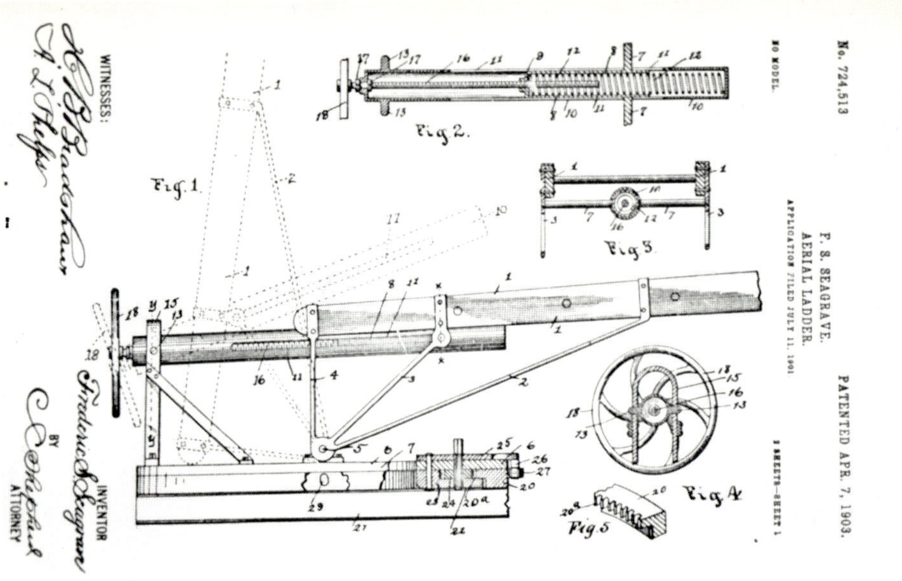

Doing research for a book has its bright moments. The photo, above, was taken while Sara Jane Sturdevant and I discussed the patent model for the Seagrave trussed ladder, shown at the left. She is the great granddaughter of Frederic Seagrave. Although she is in no way connected to the fire equipment business today, she still had numerous photos of early apparatus. Her colorful recollections of past history helped document the first chapters of this book.

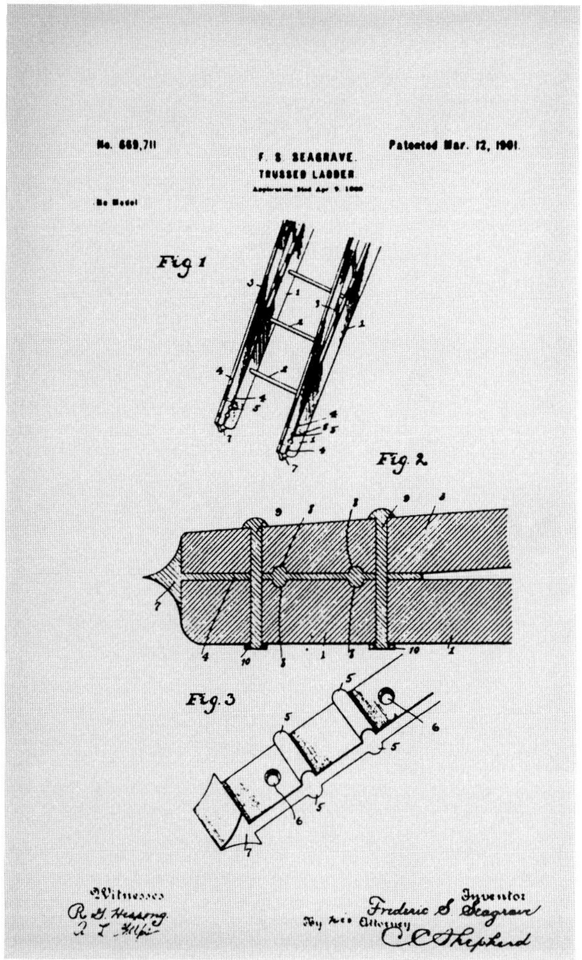

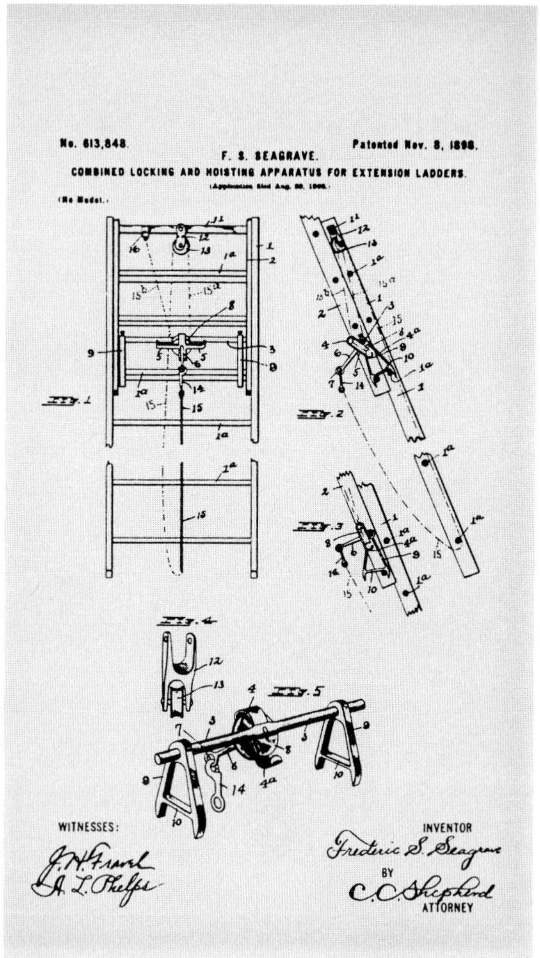

Seagrave patented the ladder items shown on these pages. All the ladders were made of clear Oregon Fir, air seasoned. Rungs were Hickory, Oak or Ash. The rungs were bolted on every fourth rung, the bolts having special nuts to avoid sharp corners. Stay poles are of the same stock as the ladder. Extensions under 30 feet are built separate so that the two pieces may be used as separate ladders. A 6 foot ladder base, available on ladders over 45 feet, folds out of the way when not used.

Size	1914 Price
30 foot extension	$ 70.00
40 foot extension	115.00
74 foot extension	325.00
20 foot wall	17.00
30 foot wall	30.00
20 foot roof	20.00
Extendable base	40.00

NOTE

Shipping ladders was cumbersome. Ladders over 35 feet, when shipped separate from trucks, took a minimum shipping weight of 4,000 lbs. Ladders under 25 feet were weighed and costs were set at a rate double their weight classification.

SEAGRAVE TRUSSED LADDERS
SHOWING EXTENSIBLE BASE AND WALL LADDERS

No. 10 Hand Hose Reel.

Intended to carry 600 feet of 2½-inch fire hose. It is a solid, well-made reel, with steel frame and hand forgings, brass-trimmed, with 52-inch Sarvern wheels. Equipped with tool-box and roller, fire axe, steel crow-bar, 60 feet of drag rope, and bracket reel; finished in vermilion striped in gold. Unquestionably the best proportioned, finished and strongest reel at the price. Weight 275 lbs., extreme width 70 inches.

No. 10 Hand Hose Reel, capacity 600 feet 2½-inch fire hose.............each, $75 00

No. 10A Hand Hose Reel.—Capacity 600 feet 2½-inch fire hose, with solid brass drag rope reel, brass pole crab and hook, weight 300 lbs.......................each, $90 00

If desired with 5½-foot wheels, add $10 00 to above prices.

No. 10A HAND HOSE REEL.

These hose reels were referred to as balanced reels because the load was centered on the axle. Seagrave hose reels all measured 9 feet 6 inches in length, and between 5 feet and 6 feet in height, but the width determined carrying capacity. A number 7 reel was 4 feet 10 inches wide and carried 350 feet of hose.

The tool or equipment box on the rear held spanner wrenches for hose attachment, a hydrant wrench and nozzle tips. The upright cone on the handle, or tongue was for storage of play pipes. The tougue rest folded up and was held out of the way by the small hook under the tongue, visible in the sketch at left.

The hand reel, number 10A, shown in this 1894 catalog sketch, rose from $90 to $190 by the time the 1914 catalog was printed.

The number 16 hand hose reel above featured 5 1/2 foot wheels, 60 feet of drag rope, three chime bells and brass trim. This 700 foot capacity reel weighed 400 pounds.

In the early 1890's, the town of Wenham, Massachusetts, like many small towns, stored ladders and water containers all over town. However, they kept the steamer and hose reel in a central location.

The 60 foot drag rope allowed many men to pull the loaded cart. Sometimes, the trick was stopping. Often, a short rope would be attached to the rear of these carts, and a lone fireman would run behind the cart, with rope in hand, to act as a brake.

Fire department hose of the day was double layers of woven cotton lined with rubber. The typical industrial use hose was not rubber lined.

Smaller carts were also made for industrial or institutional use: factories, wharves and warehouses. Some warehouse hose reels were made with wooden frames and 30 inch diameter agricultural wheels. They were less than 3 feet wide (to get through doorways), and cost less than $30. The largest of the plain warehouse reels were 54 inches wide, had a steel reinforced frame and cost $45.

Seagrave would supply its own make of soda acid extinguishers or the standard Babcock brand extinguisher. Babcock's were in the 2 1/2 and 6 gallon sizes. Seagrave furnished 3 and 5 gallon sizes. The chemical charge (soda and acid) would provide enough pressure to throw a stream 40 feet. In plain copper, the price was $25, plus $4 extra for nickel plating. The apparatus carried extra bags of bicarbonate of soda and an acid jug to recharge the extinguisher.

The weight of the apparatus determined the wheel size and the hub type. The apparatus, above, has two piece Sarven type hubs. The apparatus, at left, has Warner four piece Army type hubs.

Augusta, Georgia ordered this tiller steer ladder truck in 1906. The rollers on this wagon would help in loading the 50 foot extension ladder that weighed 300 pounds without stay poles, and 342 pounds with stay poles.

In 1902, Seagrave moved into new and larger quarters at 2000 South High Street. The new facility was the former home of The Ohio Buggy Company. The building shown here, in this artist-enhanced rendering, shows an overpass for the railroad. This overpass, and the loading dock (marked with an "x"), were to become the backdrop of many photos for the next fifty years.

The photo of this tiller rear steer ladder for Augusta, Georgia was taken on the loading dock. A further explanation of the buildings and their layout appears on page 204.

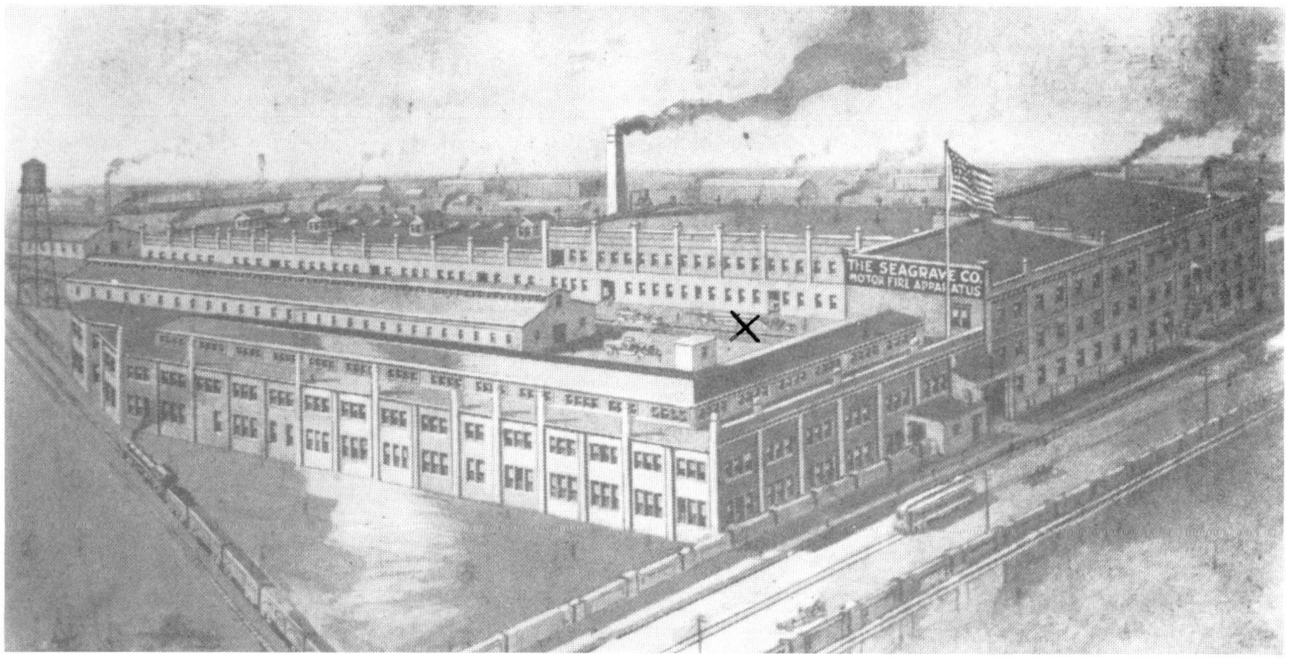

General Offices and Factory, Columbus, Ohio

The chemical tank of 40, 50, or 60 gallon capacity mounted on an apparatus became a Seagrave standard for decades. The twin tank unit, above, had all the extras, including folding supports, front and rear, to keep the tanks level during long periods of storage. The basket carried 150 feet of 3/4 inch rubber hose. Extra charges were carried in the box at the rear step. Made of solid copper, plated with nickel, and having a hydrant connection for recharging, the prices were:

Double 20 gallon tanks - $750
Double 30 gallon tanks - $850

The charge of a 30 gallon tank would be 12 pounds of soda and 6 pounds of acid. In a tank full of water, it could produce an initial surge of 400 psi, depending on the temperature at the time. Chemical hose of three ply woven cotton was rubber lined and rubber coated. The 3/4 inch size cost 55 cents a foot, in 1914.

Indianapolis, combination Hook & Ladder 6 boasted a 40 gallon tank and 150 feet of hose. The rest of the equipment consisted of 9 ladders, 2 axes, 4 pike poles, pull down hooks, 2 three gallon extinguishers, 4 lanterns, 2 torches, a crow bar and an assortment of hand tools.

The hose and chemical deliveries to San Diego and Washington D. C., featured 25 gallon tanks and the new ventilated side metal hose boxes.

75
THE SEAGRAVE COMPA
OWNERS OF SEAGRAVE'S PATENTS.
FIRE DEPARTMENT APPARATU

Some communities, like New York City, and Chicago, Illinois, received multiple deliveries. New York's "second size" wagons featured the extra rear facing seat, overhead rack for pompier ladders and a locker under the tool box for a rope style life-net. The description second size was steamer terminology.

These stout wagons, ordered in mass, became obsolete with the retirement of the horses in the 1920's. New York's last horse was sold in 1925. A second life for many fire department wagons throughout the county became the menial task of trash pick up.

Chicago deliveries of 1906 featured a separate seat for two. Chicago retained this separate seat arrangement well into the 1940's.

Many communities even ordered chief's buggys with the rest of the apparatus order. The buggy, shown at the right, was for Rochester, New York.

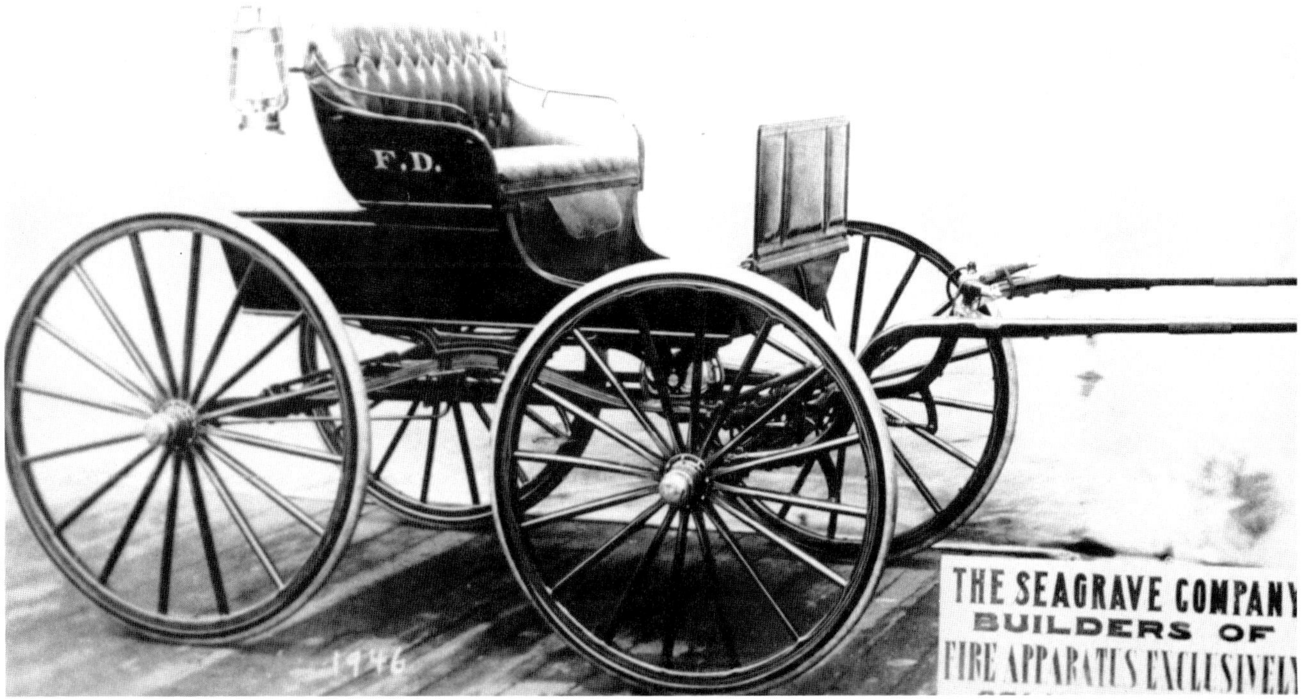

Shown on the dock, prior to shipment, is the whole Murray City, Utah Fire Department of 1906. It was not unusual for a department to order its whole fleet at one time. Murray City still has one of the hose carts. It is presently awaiting restoration.

Early multiple orders were usually received after a major fire forced the townsfolk into making the purchase.

Horse-drawn, hose reel wagons were not common in most areas of the country. Delivery of hose is faster from a reel, but the capacity was half that of a standard wagon.

The overall length of this reel was 12 feet. The front wheels were 42 inches in diameter and the rear wheels 52 inches in diameter. With a wheel base of only 58 inches, a tread of 55 inches and a top heavy load of hose, this hose cart would exhibit poor handling characteristics.

Three Rivers, Quebec, ordered this tillered service ladder truck, in 1913. With our snow removal equipment trucks of today it is hard to imagine unplowed roads.

Seagrave of Columbus, Ohio and W. E. Seagrave of Walkerville, Ontario, cornered much of the Canadian market. The Seagrave name was the most dominant until American LaFrance built a shop on Weston Road in Toronto, Ontario, in 1915. The LaFrance shop expanded, in 1916. About that same time Seagrave production was reduced by a serious fire at the Walkerville, Ontario, plant.

The fire took its toll, but loyal customers would reorder. Another delivery to Three Rivers, Quebec, is shown on page 246.

In Northern climates it is not unusual to have snow on the ground three or four months of the year. Snow could pack into the spokes of the wheels and the effect would be the same as rolling a large snowball.

THE SEAGRAVE COMPANY

Manufacturer's of High Grade Fire Apparatus

NOT IN ANY TRUST OR COMBINATION **COLUMBUS, OHIO**

The Seagrave Company's Fire Department Trussed Ladders for Aerial and Service Trucks are the maximum of strength and rigidity and the minimum weight. The ladder on trestles is 20 ft. long, 28 in. wide and weighs 68 pounds. The aggregate weight of the men, standing on every rung, except two at each end, is 1,162 pounds. The ladder leaned against the building is 18 ft. long, 18 in. wide and weighs 45 pounds. The aggregate weight of the men (one on every other rung) is 1,085 pounds.

Photograph taken and men weighed Nov. 5th, 1907, Columbus, Ohio.

Built under patent, July 19th, '98(2); November 8th, '98; May 16th, '99; January 22d, '01 (2); March 12th, '01; September 10th, '07.

The ad at left, appeared in many trade journals. The picture is truly worth a thousand words. Seagrave photographers of this period used a camera that incorporated a glass negative. The shutter of the camera would stay open several seconds depending on the light available. People in the photo were told not to move and to keep their hands folded or braced so as not to blur the photo.

Sturdy ladders and durable wagons were Seagrave's stock in trade. Seagrave stock ladders were arched on each side. Seagrave roof ladders were flat on one side so they could lay flat on a roof.

The simplicity of Seagrave designs can also be shown in the design of the "fifth wheel", or pivot for the horse hitch. A steel strapping is carrying the load, reinforcing the wood framing at the critical point. This wood and steel construction would prove so durable in framing members, that it would not be until 1936 that cabs became all steel. And, steel skinned wood doors survived into the 1960's.

S. S. M. No. 1
THE SEAGRAVE COMPANY
BUILDERS
COLUMBUS OHIO. U.S.A.
NO. 14501

LOAN CO.

SEAG

The last true wagon was built for Sault Ste. Marie, Michigan, in 1916. But, horse-drawn apparatus would be on the scene of major fires for years.

Shown, lower left, is a long exposure photo taken at a night fire in Columbus, Ohio. The trails or streaks were caused by the lanterns of firemen as they moved about the fire scene, in this long exposure photo process. (Today, we see photos of streaked tail lights from autos on a roadway that are taken by this same process.)

The aerial ladder photos show, Seagrave designers were busy with a new project, the spring raised aerial ladder. Between 1903 and 1905, Seagrave perfected the spring raised ladder system. The ladders weight was actually counter balanced by springs compressed in the tubes shown near the base of the ladder.

These aerial trucks are both rebuilds. The upper photo shows a Babcock chassis with a newly installed aerial hoist featuring the Seagrave truss beam ladder, and a low pivot point for the aerial ladder. The lower delivery, for New York, shows a Hayes aerial ladder with Seagrave ground ladders. The Hayes ladder featured a high pivot point for the aerial. Hayes was a master mechanic at the San Francisco Fire Department. He later sold this design to American LaFrance.

These deliveries are Seagrave all the way. They are from 1903-1905 with the low pivot point for the main ladder. After the photo, the loose items were boxed and shipped separately. The Hocking Valley Railroad, which served the plant, became the Thurston & Ohio Central, and finally the New York Central. These deliveries went to St. Paul, Minnesota and Salt Lake City, Utah.

By 1906, the ladder pivot base was raised off the turntable. The springs for lifting the ladder and counter balancing its weight are now mounted at the base. The rectangular pockets bolted to the top of the turntable were for insertion of rotation bars. The base of the ladder was manually rotated with poles, like a ship's capstan. The ladder itself is of the wooden truss style. A change in ladder styles paralleled the advent of the motor-drawn apparatus, as shown on page 72.

The wooden aerial ladder could support a lot of weight, but it was no match against a steel tower for containing the thrust generated by a large bore nozzle. Frederic Seagrave designed and built a water tower for Winnipeg, Manitoba, in 1905. The following year he built its twin for Montreal, Quebec, shown here.

The Seagrave design used a triangular mast (to reduce weight) and a rear mounted turntable. The rear mount was to reduce total overhang. When the 65 foot mast was lowered, the whole apparatus was less than 36 feet 6 inches.

Montreal's tower had an assignment of eight men and four horse. The photo above shows a practice run, in 1907 (No smoke from the steamers). The tower was motorized, in 1914, with a Seagrave tractor. With more modernizations and modifications it remained in service, until 1946. By then the versatility of the all steel aerial ladder doomed the single function water towers.

THE SEAGRAVE RAPID RAISING WATER TOWER.

Built Under Patents December 5, 1905-September 10, 1907.

Base of tower is attached to ball bearing turntable with steel forgings. Springs operating by expansion, balancing the weight of the tower mast are enclosed in steel tubes permanently attached to turntable, are governed by screw attached to the mast by an adjustable ball and roller bearing journal box hung on vertical axis working in a bronze nut attached to frame hung on both horizontal and vertical axis, which absolutely avoids all possibility of binding. The tower is telescopic, made in two sections; the upper section being a steel tube to which the hose is attached, is raised by drum and cable. On top of the upper section is a flexible pipe with discharge nozzle controlled by a man on the turntable, turning the stream in any direction at will. The tower is provided with the Seagrave Automatic Controlling Valve, a recent patented invention owned by the Seagrave Company, which prevents the tower turning over by the reaction of the stream of water; the valve acts automatically—the instant the wheels leave the ground the valve immediately cuts off the supply pipe at the base of the tower. The tower is also provided with a deck turret nozzle, of any style preferred, placed on the deck of the frame, and is arranged to throw a stream at any angle desired, convenient for use in lower floors. Wheels, 3-inch spokes, tired with 3 x 1-inch R. E. open hearth steel or 4-inch rubber tire, tracks 6 ft. 6 inches center to center, axles 3-inch, square bed, frame 6-inch, channel 20 feet long, reinforced. Tower pipe extends to 50, 55 or 70 feet. Water receiver of steel pipe connecting both tower and deck turret; either may be shut off by a valve at any time.

Shown, at the left, is a Seagrave rebuild of an 1895 Champion water tower for the New York City Fire Department. The catalog page, and photo above, highlight New York City's 1907 water tower, serial #2097.

The Buckboard Era

1907 - 1916

Warren Edmund Seagrave **Frederic Scott Seagrave**

W. E. Seagrave (far left) and F. S. Seagrave (far right) with the new delivery for Ottawa, Ontario, in 1911.

North America during this period was to come to the forefront in manufacturing skills and volume production. Early hand-drawn equipment had been imported from Europe. Most steam-driven equipment was hand crafted using only the skill of the craftsman to control quality and interchangeability of parts. The Buckboard Era at Seagrave incorporated the best work of the master craftsman and efficient quality control.

The early 1900's were a time of drastic change, the automobile was coming of age. To be, and stay, competitive Seagrave would have to adapt. Most trucks of the era were adaptations of autos on a more substantial chassis. This would not do for the weight and size of fire apparatus.

Seagrave management was fortunate to have one of the most dependable producers of automobiles right in its own backyard. In 1902, Lee A. Frayer and William Miller were developing engines for use in cars and trucks. The Frayer-Miller motor cars of 1904 to 1907 proved to be exceptionally well built. The first six-cylinder auto in the U. S. was an air-cooled 1905 Frayer-Miller of 36 horsepower.

The year 1906 brought much glory as three special built four-cylinder cars were among the finishers of the Vanderbilt Cup races that year. In fact, Frayer-Miller cars finished all races they entered. Design and race testing, however, took its financial toll. In 1907, Frayer-Miller joined with a local entrepreneur, Oscar Lear, a promoter and seller of new gadgets. Lear approached the Seagrave management about sales of its commercial line. The standard Frayer-Miller chassis proved too light for fire service. But, the engines performed well, and Seagrave purchased engines in many sizes, through 1909.

Financial problems plagued Frayer-Miller. Seagrave purchased manufacturing rights to the larger engines and much tooling at a bankruptcy sale, in 1909. With an excellent power source, and Seagrave know how in apparatus production, a new era was created.

The photo at left is interesting because it shows F. S. Seagrave is now slightly apart from the business he created. The photo also foreshadows the advent of cab forward fire trucks and personalized license plates.

THE SEAGRAVE COMPANY
COLUMBUS OHIO
CHAS. F. KIESER SUPT.
THE SEAGRAVE CO.
CHIEF FILLMORE TYSON
LOUISVILLE KY.
CHIEF SAM'L F. HUNTER
SPRINGFIELD O.
HON. C. A. BURKHARDT
CHIEF FIRE DEPT
ST. BERNARD O.
CHIEF J. F. MERTZ
AKRON O.
C. W. BLACK TREAS.
A. G. CROUSE GEN'L AGT.
THE SEAGRAVE CO.
PHOTOGRAPH OF PARTY ON RETURN FROM 55 MILE RUN FROM
CHILLICOTHE TO COLUMBUS JUNE 27TH 1907.
PHOTO. BY B. K. BLACK.
THE SEAGRAVE CO.

The first Seagrave motor vehicle made news, and the photo at the left, was published widely. The whole trip was 105 miles and took eight hours and 17 minutes. The average speed was 13 mph.

This unit was a four-cylinder, 24 hp model on a Frayer-Miller chassis. Axles were Timken, and it was fitted with Swinehart Clincher tires.

The photo at the right, shows the chemical and hose wagon Chicago tested for 30 days, in 1908. They found it too light for service and did not use any motorized equipment until 1911.

Refinements after these in service tests led to a larger clutch, more air flow for cooling, and lengthening of the wheel base for more carrying capacity. Frayer-Miller ads of the day boasted of "60 to 90 mile trips completed in a single day", and "your air cooled truck is available to work every day of the year, no radiator freeze ups or radiator boil overs."

24-H.P. FRAYER-MILLER STAKE TRUCK. Capacity, 3,000 lbs.

FRAYER-MILLER Motor Truck

At the 1909 bankruptcy, Seagrave purchased manufacturing rights to some Frayer-Miller components and some of the stock of the existing automobiles. The remainder of the concern became part of Kelly-Springfield. Kelly trucks were produced through 1928.

The first Seagrave motor apparatus were on Frayer-Miller chassis. The single window, for motor access, denotes a Frayer-Miller chassis.

The first motorized aerial ladder truck

Seagrave's main drawing card for the Fire Engineer's Convention, of 1908, was this tractor-drawn aerial. The special tractor, fitted to the standard aerial, featured a cold-rolled steel frame and an air cooled four-cylinder engine.

The following statistics prove the massiveness of this machine. The wheel base of the tractor was 130 inches. The 40 inch tires were four inches wide. External brake pads were applied with the foot brake, and internal brake shoes by the hand lever. The front axle is a 2-1/2 inch "I" beam and the rear is a solid three inch square. Tractor and trailer spread is 66 inches. The distance between the tractor and trailer axles was 22 feet, 9 inches, and overall length was 53 feet.

Motor fire apparatus was an easy sell to a progressive fire chief. The motor apparatus cost more to purchase, but after the initial purchase, daily costs dropped. Unlike horses, the motor apparatus could be fresh for even six fire runs in a row. The speedy response time, coupled with increased load carrying capacity lead to rapid adaptation of motorized equipment. Fire station locations could be farther apart and still provide adequate coverage. Less time was also needed on horse care and harness repair.

Another point was the driver of a horse-drawn wagon was out of service if there was no fire buff at the scene to tend the horses.

The upper photos on this page are of the very first Seagrave motor fire apparatus sold. It went to Vancouver, British Columbia, in 1907, and it carried factory serial #2382. Vancouver Chief J. H. Carlisle was so pleased with the performance that in 1909, he ordered again. The 1909 deliveries included two wagons and an aerial ladder. They are shown at the right and in the group photo above. A third order was placed for three chemical cars and two more chemical wagons in 1910. The repeat sales made a statement about the quality and reliability of Seagrave apparatus.

V. F. D.
THE SEAGRAVE COMPANY
BUILDERS
COLUMBUS OHIO, U.S.A.
No. 8301

3301

THE SEAGRAVE COMPANY

The police wagon for Gary, Indiana, was built, in 1907. It was the prototype vehicle for many design changes. Early changes were the balloon tires, developed by John Dunlop in 1902, which proved to be unstable under a fully loaded condition. The rear fenders proved insufficient as the chain and drive sprockets also were 'mud slingers'. The engine service area was also inadequate. The police wagon topped the scales at over 4,000 pounds.

Differences between serial #3823, a 1909 delivery for Wheeling, West Virginia, and serial #5980, a 1911 delivery for Denver, Colorado, indicate another set of design changes. The front fenders became enclosed and the rear fenders more form fitting. A step was also added for ease of entrance. The parking brake lever was moved and the steering wheel angle became more driver oriented.

Seagrave apparatus was available in five motor sizes (see page 89). Wheel bases of 112, 122 and 130 inches were standard.

The hood up on this tractor reveals the type E motor. Valves are located on either side of the top of the cylinders. Dual spark plugs were standard. One ignition system was a distributor and dry cell type and the other a magneto. Power is transmitted through a leather cone clutch. The transmission had four forward speeds, direct in fourth, and a reverse gear. Gear ratios were designed to allow the rig, fully equipped, to climb a 27° grade at 25 mph. In actual tests, the apparatus had made a three mile run and the ladder was put in service in less than six minutes.

W.W.F.D.No
THE SEAGRAVE COMPANY
BUILDERS
COLUMBUS OHIO. U.S.A.
NO. 3748

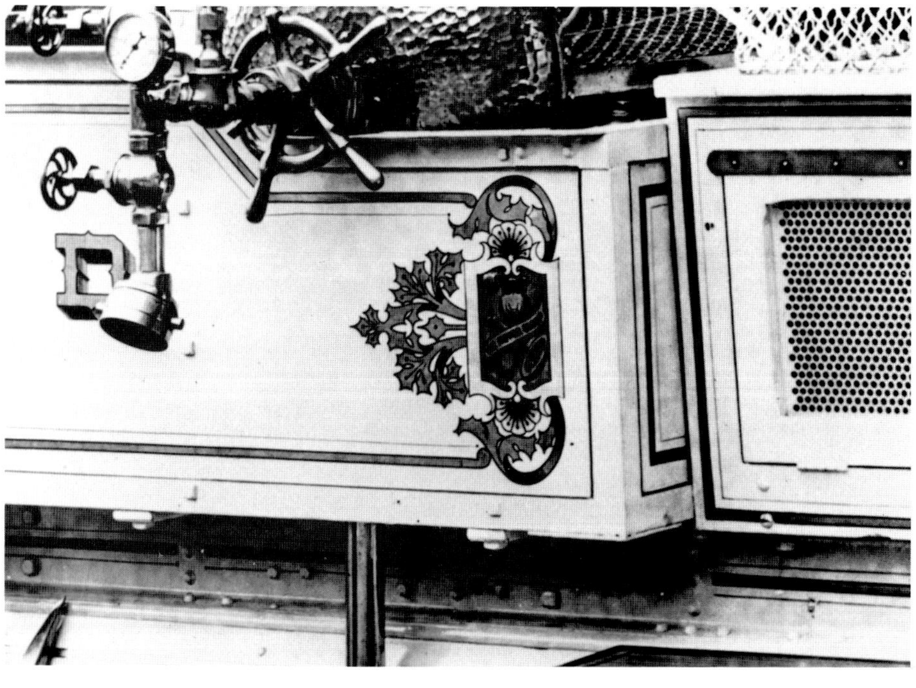

Decorations

Fine painting and striping graced the majority of early Seagrave apparatus. Bright trim of real hammered gold leaf adorned not only fire equipment, but most fine articles of the day; gilded edge books to private railcars. Fine designs on leather table tops and trim on the cover of the family bible all denoted craftsmanship. The total weight of gold for a complete job was less than three ounces. It's tedious to apply, but the shine is eternal.

Walla Walla, Washington's, serial #3748 was painted in December, 1909; Redlands, California's, serial #4324 in June, 1910; and Tacoma, Washington's, in August, 1910. Redlands type C four-cylinder used a D-4 Bosch magneto and an Atwater Kent coil system. The carburetor was a Stromberg B-7.

The type D six-cylinder applications used by Walla Walla and Redlands were easy to spot by the enlarged side grille openings. Tacoma received the chemical, shown above, and an aerial tractor, shown at left and on page 75.

CHEMICAL TANK AND FIRE EXTINGUISHER DEPARTMENT.

FOUNDRY.

MOTOR ASSEMBLY DEPARTMENT.

CORNER IN MOTOR ASSEMBLY DEPARTMENT.

The main building of Seagrave was built for the manufacturing of buggies, in 1895. The relatively small size of the buggies and their light weight led to a facility with low ceilings and light flooring. Buggy and wagon manufacturers were plagued with storing the equipment while the paint and varnish dried. This same dilemma was partially solved by Henry Ford, in later years, by only offering black, which absorbed light and dried faster.

REAR WHEEL BRAKE TESTING DEPARTMENT.

VIEW IN ONE PAINT SHOP.

FINISHING, PAINT AND DECORATING DEPARTMENT.

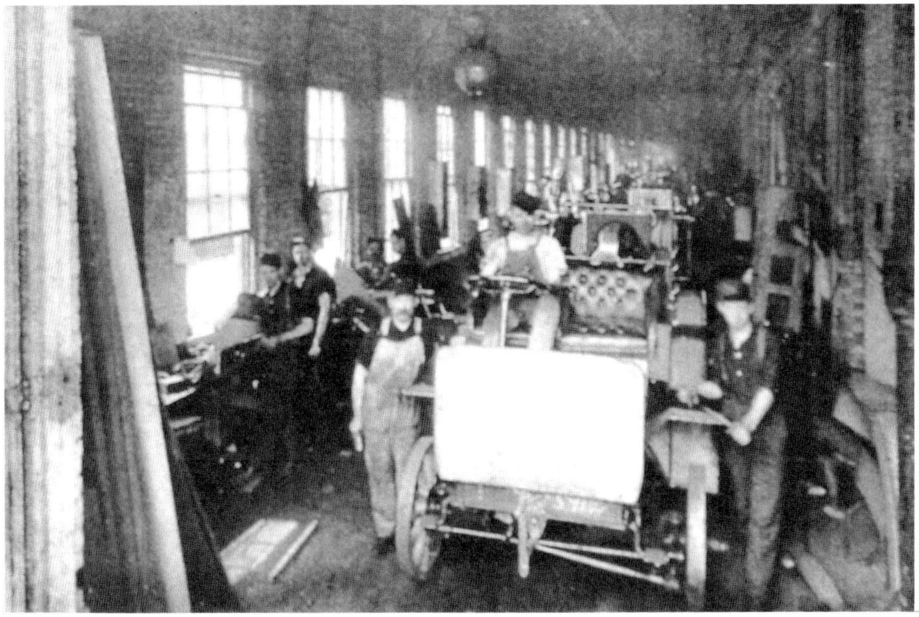

FENDER AND BODY SHOP.

By the time these photos were taken, in 1911, Seagrave employed 180 people. Total production for the year was 120 rigs. Horse-drawn wagons accounted for less than 15 percent of the total. Rebuilds and modernization of earlier equipment were made in the blacksmith shop. Then, as now, rebuilding and updating equipment is a vital service of apparatus builders.

The photos on these two pages depict two of the buckboard style survivors. Columbus, Nebraska, and Pasadena, California, still have their original deliveries.

Needing minor repairs and adjustments, in 1912, the factory sent the Columbus, Nebraska, fire department some blueprints. These prints, dated through 1911, show how Seagrave updated and strengthened the earlier Frayer-Miller design. By 1911, all prints carried the Seagrave name. Columbus' rig remains unrestored and very original, except for the 1920's addition of electric lights.

Note the photographer's assistant in the photo, at left. With a long shutter opening, the helper's job was to shape the back drape so it would appear as a solid white background.

The sign in the upper photo shows gasoline at the lowest market. In those days, gasoline in rural communities was available in hardware stores and was often used as a spot remover or degreaser.

Pasadena, California's, Seagrave has been the star of many Tournament of Rose parades. But, life wasn't always that glamorous. On January 1, 1910, the new Seagrave rode in its first parade. The original price for the 40 hp apparatus was $4,950. After 13 years service as a chemical wagon, it was reworked into a search light car. In 1936, the rig was retired and stored sometimes inside, sometimes outside until 1964. Restoration by the department and local historians returned chemical engine No. 5 to her original splendor in time for the Tournament of Roses Parade, in 1970. Cost of the restoration was $3,600 and over 4,000 man hours.

SIDE VIEW OF TYPE D CHASSIS.

The type D chassis featured designs available only on the finest 'high grade' trucks of the day. Timken drop forged axles and bearings, Vanadium steel springs, double ignition on two sets of spark plugs, and Raybestos brake linings were standard equipment.

All deliveries, after 1908, had a pin to hold the hand crank up and out of the way when not in use. Deliveries, after 1912, had the hand brake and shift lever moved to the drivers left. The change allowed for easier entry to the driver's stand.

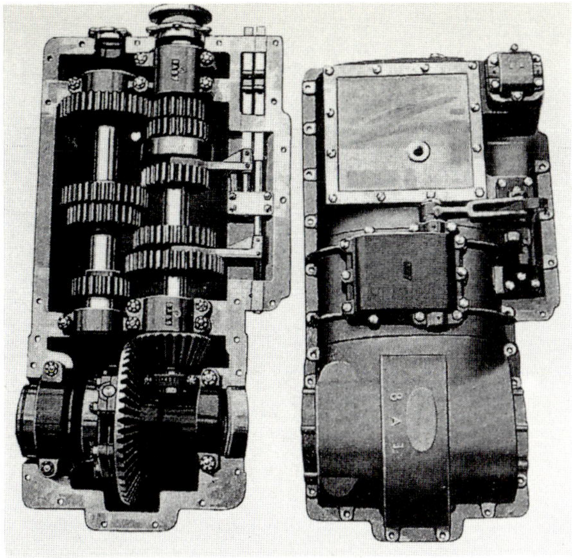

Transmission and Jack Shaft

The transmission and jack shaft are enclosed in one housing cast in two parts. The lower section is bolted to frame and carries all bearings, shafts, gears, etc., and the upper section is merely a cover and can be removed at any time for inspection. The transmission is selective type, having three speeds forward and one reverse, direct drive on third speed. The reverse gear is automatically locked in reverse and neutral position. All gears run in oil bath. All shafts are mounted on Hess-Bright ball bearings. All gears and shafts are made of the highest quality special automobile clash gear steel specially heat treated and hardened. All gears are of the stub tooth type, giving an increase of 35 per cent in efficiency.

Los Angeles, California, received this chemical wagon, in 1911. It was built on the D chassis. After a few years of service, the department added a Gorter deck gun to the equipment carried by its 'Great White Hope'. Henry A. Gorter, the guns developer, was the superintendent of engines for the San Francisco, California, Fire Department.

The transmission above powered a separate driving axle called a jack shaft. Having a jack shaft removed torsional stress and strain from the rear axle, which then only had to carry the load. The use of a jack shaft system allowed for more ground clearance and a dual brake system. The main brake, the foot brake, was a contracting band on the jack shaft drum.

The West was Seagrave territory. The top photo shows Chief Harry Meyers of the Spokane, Washington, Fire Department, with his new Cadillac Chief's Buggy and his 1911 Seagrave chemical wagons.
The lower photo shows the Victoria, British Columbia, Fire Department, in 1913, with its Seagrave fleet.

Above is a March, 1913, delivery to Richmond, California. The photo, at right, is a 1911 delivery to San Diego, California.

Compare these rigs to the 1909 deliveries to Vancouver (see page 55), they are all just two years apart. It's interesting to note the differences. The 1909 deliveries have chain guards, eight inch acetylene headlights and a flywheel driven siren.

The 1911 deliveries had running board splash shields, a slanted steering column and a bumper. By 1913, the headlights were electric, the fenders were full form, and the oval gas tank made its first appearance. The oval gas tank was a feature through 1937.

L.F.D.
THE SEAGRAVE COMPANY
BUILDERS
COLUMBUS OHIO. U.S.A.

Y.F.D.
No. 9
THE SEAGRAVE COMPANY
BUILDERS
COLUMBUS OHIO. U.S.A.
No. 4405

G.F.D.
THE SEAGRAVE COMPANY
BUILDERS
COLUMBUS OHIO. U.S.A.

REDDING FIRE DEPT.
THE SEAGRAVE COMPANY
BUILDERS
COLUMBUS OHIO. U.S.A.

B.F.D. No. 14
THE SEAGRAVE COMPANY
BUILDERS
COLUMBUS OHIO. U.S.A.

BUILDERS

Photos at Left

A diversified custom builder can provide almost any configuration, as indicated by the variations of the standard buckboard theme.

Serial #	Destination
3571	Loveland, Colorado
4223	Greeley, Colorado
4405	Youngstown, Ohio
4501	Birmingham, Alabama
5369	Reading, California
5431	Grand Junction, Colorado

Photos at right

Grand Rapids, Michigan, ordered this modern version Seagrave in December, 1911. The apparatus was the first Seagrave with a water cooled engine. Seagrave joined the trend away from air cooled engines because of overheating conditions. Also trend setting was the use of air tires on the front end for ride control. The rear axle remained on solid tires because of their exceptional load carrying capacity.

Today, we would call this a "cab and chassis". When Los Angeles, California, ordered this rig in 1911, it was referred as an "operator's stand and chassis".

Many departments preferred to transfer a horse-drawn hose wagon body to the new motor chassis in an effort to save money. On many pieces of equipment, the motive power system didn't really wear out, it just became obsolete. On these early deliveries, equipment would be transferred from one chassis to another to take advantage of more horsepower or less temperamental drivelines.

Chief John Aitken of the London, Ontario, Fire Department inspected every motor apparatus he could find. He also wrote to chiefs of departments using motor apparatus. Replies like the following from New Orleans, Louisiana, dated September 5, 1910, convinced him to go Seagrave. ". . . We have placed a double tank Seagrave in service June 1, 1909, and it has responded to three alarms daily on average since then. Total maintenance costs have not exceeded $1.25 to the present time."

The London Fire Department received their delivery, in 1911. In June, 1913, it was hit by a train at the Williams Street crossing. The apparatus was sent back to the factory for rebuilding.

W. E. SEAGRAVE
F. S. SEAGRAVE

W. E. SEAGRAVE & CO.

Originators of

SEAGRAVE FIRE APPARATUS

IN THE UNITED STATES AND CANADA

"We Claim"

Greater Efficiency
Larger Capacity
Lower Cost of Maintenance

OFFICE AND FACTORY
WALKERVILLE, ONTARIO

The photo, at right, shows F. S. Seagrave riding on the hood of a recent pumping engine to Windsor, Ontario. The testing of this apparatus is shown on page 96.

When Frederic Seagrave sold his interest in the manufacture of fire apparatus, in 1898, to Julius Stone and a group of Columbus, Ohio, investors, he kept the right to produce and sell under the Seagrave name in Canada.

Warren Edmund Seagrave, Frederic's son, directed the operation. He had been sales manager for the U. S. organization.

Headquarters were set up in Walkerville, Ontario. Walkerville, now part of Windsor, just across from Detroit, was an ideal location. Windsor and Detroit were both rail heads. Railroad car ferries traversed the river continually. A mile long railroad tunnel under the river was completed, in 1910, to further enhance doing business in Windsor. Soon after establishing W. E. Seagrave & Co., Frederic went into semi-retirement again.

W. E. Seagrave & Co. received partially completed apparatus from Columbus, provided locally built bodies, and sold them throughout Canada. The customs duty on complete fire apparatus was as high as 35 percent, in 1904.

W. E. Seagrave entered the commercial vehicle field in 1917, with trucks of a capacity up to 3 1/2 tons. In 1919, W. E. Seagrave & Co. merged with Loughead Machine Company of nearby Sarnia, Ontario. An economic slow period caused the demise of the company, in 1923.

Commercial Seagrave deliveries were powered with engines from Waukesha and Herschell-Spillman. In 1917, Seagrave produced the first V-8 powered truck in North America.

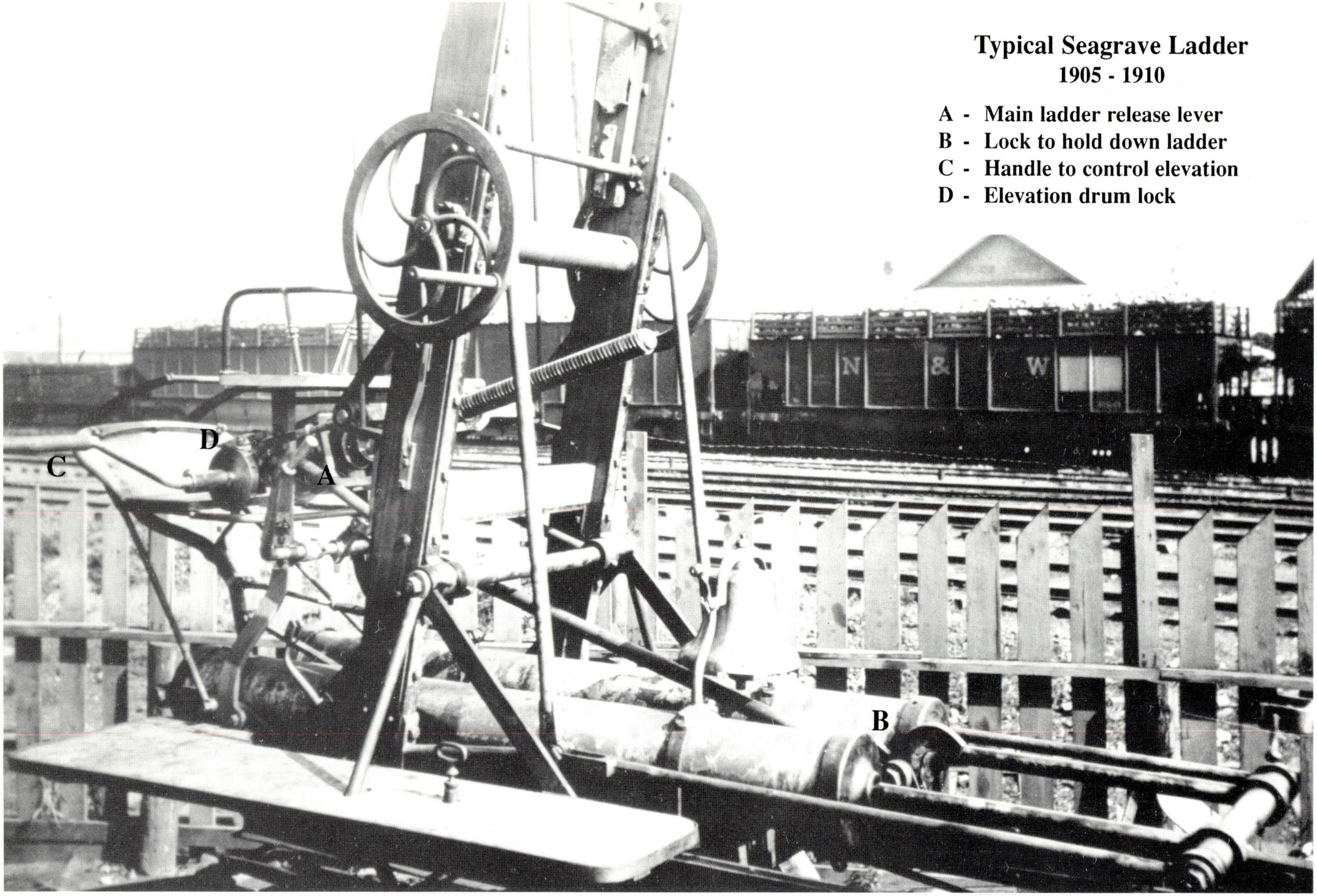

Main aerial ladders of the late horse-drawn era and the early motor era were made of Oregon Fir. The factory stocked straight grain ladder stock in up to 40 foot lengths. These straight side ladders were trussed with steel rods, supported every fourth rung. The steel rods carried the principal load. The five foot steel turntable, mounted on ball bearings, could be rotated by one man without the use of capstan poles. The oval handle on the operator's stand was a lock pin to hold the turntable in the nesting position. The operator's seat stand formed an easy entry to the ladderway when the ladder was raised (shielding the mechanism).

Motorized ladder deliveries featured the new straight beam ladder. Owing to their great weight, all ladder trucks were powered by large four-cylinder, 90 horsepower engines. The top photo is of San Diego, California's 1910 delivery. Spokane, Washington, had their earlier horse-drawn ladder repowered in 1911. They asked for the return of the old horse-drawn front section — "Can't trust those motors all the time". Detroit, Michigan, also asked for a repower of their 1907 65' ladder, in 1912.

Just inside the door (above) were the friendly salesmen of Seagrave, waiting to extol the joys of tractor-drawn aerial trucks.

These were huge machines designed to pull loads of twelve tons. Stopping and turning these monsters was not an easy task. Starting by hand, an engine of 1,414 c. i. displacement, was truly not for the faint hearted.

Always pull on a crank handle, never push! If the engine backfires and you're pushing, the reverse thrust could break your arm. If you're pulling up, the engine backfire would just pull the crank handle out of your hand.

The photo, at left, shows the hitch mechanism of Seattle, Washington's, 1911 delivery. Comparisons can be made with older and newer styles on pages 41 and 190.

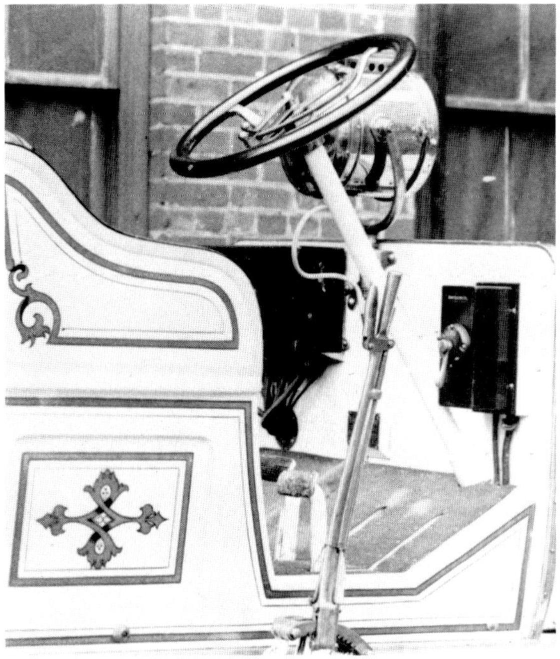

The photo, above, reveals the driver's stand, as it was called, looking very plain. Just two pedals, clutch and brake. Spark (engine timing) and gas were controlled by levers in the center of the steering wheel. The term "driver's stand" was a carryover from street car terminology.

The upper right photo shows Cleveland, Ohio's, 1913 Seagrave tractor pulling a 1908 American LaFrance ladder. The big Seagrave was Ladder 1 until it was replaced with a new Seagrave tractor-drawn aerial ladder, in 1926.

In large cities, with street cars, negotiating the tracks was often difficult. The narrow, hard rubber tires would often slip into the groove next to the rail. Tacoma, Washington's, crew is shown backing over the tracks and into their quarters.

The round hood Seagrave was made only one year, 1911. The hose and chemical for Ardmore, Oklahoma, was an air cooled 53 hp vehicle. The tractor for Los Angeles, California, was a model AC80, or air cooled 80 hp job. The round hood did not lend itself to the water cooled radiator configuration that was becoming so popular.

The product mix from the Columbus plant expanded more, in 1911. That year a touring car style fire engine became available. The change brought with it a foot pedal type accelerator, speedometer and optional clock.

The engine of the day was a dual ignition model with spark supplied by dry cell batteries or magneto. The oil supply to the motor was from an internal pump or optional externally mounted "handcock oiler" which supplied oil from a reservoir to eight lubrication points on the engine.

A Stromberg carburetor supplied the fuel and the whole engine was mounted in a subframe to isolate it from the twisting of the main frame. The main frame was a five-inch high almost two-inch wide channel, the engine subframe a three-inch channel.

After chassis and body assembly, the apparatus was taken for a trial run. Shadesville, Ohio, was seven miles south of the plant down High Street or U. S. 23. Many pieces of apparatus made their first run to Shadesville on mostly unpaved roads. During the trip, the serviceman would note driveability problems, things that should be corrected at the plant, prior to rail shipment.

The photo above shows a Vancouver, British Columbia, tractor fresh from paint, but mounted with slave wheels, for the run to Shadesville. Before loading in the box car, the slave wheels were exchanged for the correctly striped wheels.

Rome, Georgia, took delivery of the top of the line, touring car style, chemical and hose wagon, in 1911. The new styling included some new mechanical features, as well. The most important was the incorporation of the interchangeable drive sprockets. This feature allowed a chief to determine, after delivery, if he wanted more speed or more power. A telegram to the plant would send back an appropriate set of drive sprockets, for field installation.

Florence, South Carolina, received this 1911, model C, hose and chemical, in December. This apparatus, frame serial #194, was a four-cylinder, sporting a Connecticut ignition and a Bosch magneto. Other shelf items were a Timken 163 front axle, Rayfield carburetor and Fisk tires. The ten inch headlights were standard Rushmore acetyline, but the nine inch spotlight was a combination gas-electric. The motor was rebuilt in 1919, and served Florence until 1936, when it was sold to Pamplico, South Carolina. There, it served many more years.

The pride of Canton, Ohio, was this marathon grey hose and chemical car, serial #7879. The painters were allowed much latitude; compare the Canton, Ohio, and Florence, South Carolina, rigs.

The new hood shape shown on these pages was attributable to Seagrave purchasing portions of the bankrupt Frayer-Miller Company, in 1909. At the sale, Seagrave bought the rights and tooling for engines, some metal forming equipment and a stock of unsold cars.

The metal forming fixtures soon were back in service, making Seagrave hoods. The remaining stock cars were retrimmed and sold as Seagrave chief's cars.

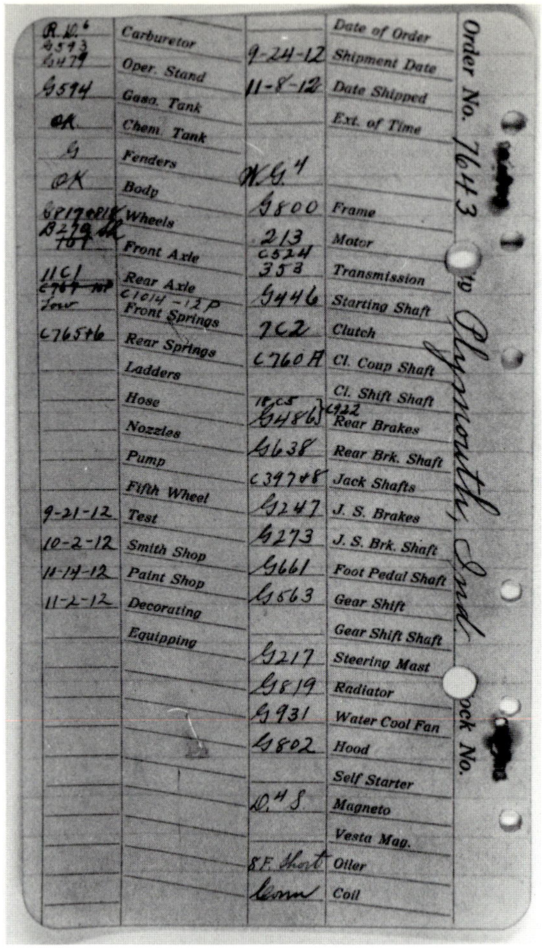

The final check sheet, above, was stapled to the back cover of every work order. It took the Smith Shop a little over a week to assemble the truck after motor approval. Then it took a month to paint and decorate.

Plymouth, Indiana, and Milwaukee, Wisconsin, received their chemical wagons only months apart. Milwaukee, home of Harley-Davidson motorcycles, ordered the six-cylinder engine and a 60 gallon chemical tank under the seat. The variation of hood lengths between these two rigs shows why parts proliferation is inherent in custom manufacturing operations.

Tractors to modernize horse-drawn equipment would constitute thirty percent of the Columbus deliveries, during the teens.

The extra weight of an aerial trailer or steamer called for driveline changes. The flywheel weight and width increased. The increased width gave more surface area for the cone type clutch to engage. The interchangeable drive sprockets were selected to provide pulling power over speed.

Sales folders extolled the value of four wheel tractors, as they could be adapted to most any steamer or aerial in an hour or so. The tractor could be moved to another piece of equipment if service was required on the steamer. The tractor after out living the steamers could be converted to a hose wagon, with a kit from the factory, for $800.

The double triangle symbol

In 1911, Seagrave General Manager, Howard Spain, started looking for a design to be placed on the hub caps and other dominant places on the vehicle. The design had to be simple, yet striking, and easily reproducible. The result: two triangles, one smaller than the other. The design appeared in mid 1911, and was used through mid 1951. First painted on the radiator shell, then as part of the emblem used through the full grille style apparatus.

Ladder trucks were the main stock in trade of Seagrave. Sales were brisk and destinations were scattered, as the photos on these pages indicate. Haverhill, Massachusetts, ordered a service ladder and chemical hose wagon, in 1913. The same year, Montreal, Quebec, ordered a tractor-drawn service ladder truck.

Seagrave truss style ladders were lighter in weight than the competition's ladders. This light weight feature allowed the Seagrave ladder truck to be lighter and more responsive than the competing makes.

Shown are three of the basic style ladder trucks, in the early teens. Top: Serial #9289, a straight frame chemical ladder for Lambertville, New Jersey. Middle: A tillered service truck for Manila, The Philippines. Bottom: Serial #9173, an 85 foot aerial ladder for Salt Lake City, Utah. All rigs were 1913 deliveries. Note the delivery for Manila is of the new shaft drive style, shown on page 85.

Seagrave experimented with this novel side saddle style aerial ladder truck, designated the model M, in both battery electric and gasoline powered versions.

The Couple Gear Freight Wheel Company, of Grand Rapids, Michigan, made electric drive vehicles, and supplied components to other manufacturers. The 1910 electric model, above, was for Springfield, Massachusetts. It, and the 1912 model for Springfield, used the five ton capacity Couple Gear wheels. The motor was in the wheel itself, and the batteries were slung under the frame.

Sales literature boasted 5 mph on a 15% grade or 25 mph on a level street, in 30° below weather. What the ad didn't state was they came out of warm quarters.

After hours of fire service time, in very cold weather, the truck could barely make it back to quarters. Only three all electrics were built, two for Springfield, Massachusetts, and one for Grand Rapids, Michigan.

Left, is serial #14164. It was an 85 foot, six-cylinder, gasoline powered, side saddle aerial for Oakland, California, delivered in 1916.

All side steer rigs were all-wheel drive, all-wheel steer. The all wheel steer feature made for easy maneuverability. But, perfect coordination was required between tillerman and driver.

On a straight frame vehicle, the tillerman can actually point the whole frame in any direction. With the separate tractor trailer, on the other hand, the tractor driver cannot be pushed off course by the tiller operator. This style apparatus, as well as the model J, straight frame service ladder, lost favor in the late teens.

The last side steer aerial went to Santa Barbara, California, in July, 1919. The aerial ladder, on the side steer chassis, was available in 65 foot, 75 foot and 85 foot sizes.

CORTLANDT HOOK & LADDER Co. No. 1

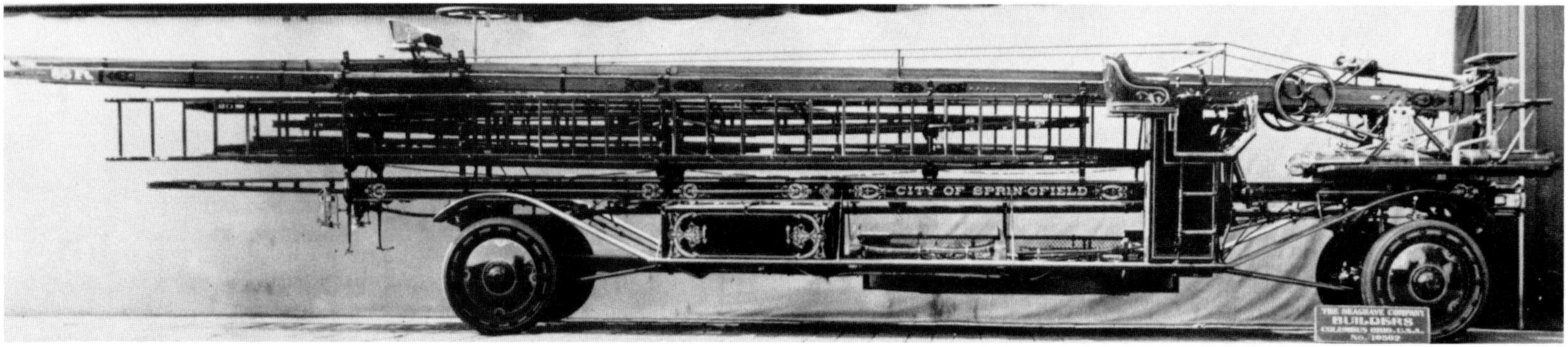
CITY OF SPRINGFIELD

Shown left are:

Top: Serial #14054, a 75 foot model for Peekskill, New York, in 1915.

Middle: Serial #10520, an all electric model for Springfield, Massachusetts, in June, 1914.

Bottom: Comparing Springfield's electric with Ladder 8, a 1916 gas engine job for Boston.

Shown above, is the main fire station for Sandusky, Ohio. The side saddle aerial ladder truck is a 1914 delivery, serial #10337. The 75 foot aerial serviced the town so well that when the engine gave out, in 1930, Sandusky ordered a new Seagrave.

The utility pole full of wires is very typical in photos of main fire stations. The phone, plus the responses for the box alarm system, usually terminated at the headquarters building. Utilities went underground in most downtown areas as part of depression era work projects.

Oh . . . those other rigs are: an American LaFrance, an Ahrens-Fox, and a Ford.

Choices galore! Seagrave produced gas engines in both four and six-cylinder versions. Each engine size was available as either water or air cooled. Also built during this early period, were all electric drive vehicles. Seagrave produced flat front, round or gable style hoods, and chassis configurations that were endless. In 1911, the year these photos were taken, the first pumper had not even been built.

During the early days, the hoods had a longer than usual appearance. The blower on the air cooled models was responsible for the extra space. Hood length, for an air or water cooled apparatus, remained the same. Some historians are often misled by the factory's insistence to put radiator filler caps on some of the air cooled apparatus deliveries.

A close up look into the shiny, silver coated reflector of serial #4815 reveals the photographer plying his trade. The white ladder tractor was for Passaic, New Jersey, and the battleship grey hose wagon was serial #6704, for Marion, Ohio. Light colors were favored by early builders as surface defects in metal forming were easier to hide using light colors.

TYPE "C" 4-CYLINDER AIR COOLED MOTOR—INTAKE SIDE.
This Motor is Made in 4 and 6 Cylinder Types.

TYPE "D" 6-CYLINDER WATER COOLED MOTOR—INTAKE SIDE.
This Motor is Made in 4 and 6 Cylinder Types.

Proven standard components, plus Seagrave engineering, provided a very stable combination. Standard part suppliers included Bosch and Atwater Kent, for ignitions. Rayfield and Stromberg supplied the carburetors. Seagrave tried to accommodate the wishes of local chiefs who preferred one brand of component over another. It was common, in those days, to specify particular brands.

On air cooled engines, a turbine blower, mounted in front of the engine, forced air through an upper air duct. This allowed the air to cool the valves, and then the rest of the cylinder. The grey iron cylinder castings were attached to a brass crank case.

All four-cylinder motors had five main bearings. All six-cylinder motors had seven main bearings. An oil pump, driven by the cam shaft, provided positive lubrication to the main bearings. Other bearings were splash lubricated. Motor designations were:

A	four-cylinder	4 1/2" bore,	5" stroke	30 hp
B	six-cylinder	4 1/2" bore,	5" stroke	40 hp
C	four-cylinder	5 3/4" bore,	6" stroke	53 hp
D	six-cylinder	5 3/4" bore,	6" stroke	80 hp
E	four-cylinder	7 1/2" bore,	8" stroke	90 hp

Air cooled cylinders were mounted on the crank case individually. Water cooled cylinders were mounted in pairs. Early catalogs would refer to a model as a WC-53. The designation meant a water cooled engine of 53 horsepower.

Seagrave never ran an iron foundry, only a brass and aluminum foundry. So, a local iron works produced the castings, which Seagrave machined.

The first centrifugal style pumping engine in America

The Gorham Fire Apparatus Company of Oakland, California, and Seagrave, joined forces to build the first motorized centrifugal pump fire apparatus. Gorham was the Seagrave west coast representative, and accounted for 40 percent of the early motorized apparatus sales. Gorham's parent company, Gorham Engineering, was a maker of centrifugal (turbine style) pumps, and the motors to drive them. They were used in irrigation, hydraulic mining and dredging. In 1911, Seagrave and Gorham joined forces to produce the apparatus, shown above. The chassis and body were a Seagrave design, and the motor and pump, a Gorham design.

It was reported, on July 27, 1912, that the first successful test took place on the Franklin Street Wharf, in Oakland. The Gorham Seagrave pumped 850 gallons per minute, for over three hours. Powered by Gorham's 7 3/4 x 9 inch six-cylinder engine, the machine had a top speed of 50 mph, and 25 mph on a 20 percent grade. Serial #6998 was delivered to the Oakland Fire Department, in August, 1912.

The first sixteen deliveries were assembled in Columbus, Ohio, minus the engines and pumps. They were then shipped to Oakland for completion.

San Diego's, 1913 delivery, shown here on the dock at Columbus, is minus engine and pump (look behind the front wheels in the above photo). Note that the front axle has been moved forward to shift some of the weight of this enormous 2,547 cu. in. or 41 liter engine.

Pasadena's delivery, right, was the second Gorham Seagrave. This delivery photo shows a long front overhang of the motor. This is typical of the very first Gorham engineered units.

The first two deliveries featured four stage pumps. Later deliveries were three stage pumps. Equipped with 40 x 5 inch front tires, and dual 40 x 4 inch rear tires, these rigs weighed 13,000 pounds.

The cross-engineering in these, and subsequent early deliveries, is significant. In 1911, when engineering plans were formulated, Seagrave did not build an engine large enough to power a 1,000 gpm pumper. Through this joint venture, Seagrave gained a desirable engine layout and working knowledge of centrifugal pumps.

Gorham could now offer its west coast buyers a full line of ladder trucks, chemicals and pumpers. By 1913, there were more Seagrave (15) motor pumpers in California, than all other makes combined. Apparatus, completed in California, carried Gorham hub caps. Columbus builds had traditional hub caps and W. E. Seagrave deliveries carried plain hub caps.

The plain hub caps indicate that this was a W. E. Seagrave delivery. Windsor, Ontario, and Regina, Saskatchewan, had four-cylinder pumping engines of 96 hp, that were officially designated as a model Q's. After this photo was taken, the "bare bones" were shipped to W. E. Seagrave for the addition of the locally made skin and trim.

The customs duty on manufactured goods was meant to keep the import price high so that local industry could be easily established. The rate fluctuated with political whim.

Right, is a photo of Columbus, Ohio's, 144 hp model. All deliveries had 6 inch rear intakes and 2-1/2 or 3 inch discharges. The all bronze fire pump and six-cylinder engine was rated at 1,100 gpm, or the old steamer designation of extra first size.

The 6th, 9th, 10th, 11th, 13th, and 14th Gorhams were delivered to the Los Angeles Fire Department. All were hose and pump combinations, except Engine 28. It was a straight pumping engine. This pumper served in front line service, until 1928. A sharp eye will note the changes that 15 years of front line duty and time bring. The five other L. A. deliveries were rebuilt into straight pumping engines, and their motors replaced with F-6 motors. Engine 28 was serial #8434.

From the first Gorham, delivered in 1912, to the last, delivered to San Diego in 1915, the centrifugal pumper set new standards.

For the first time in America, fire apparatus could pump large volumes of water with very little shake or vibration. Wear was cut to a minimum. In fact, volume often rose as the waterways became polished by the swirling action of the water flow over time.

Colton, California, received this white triple combination (pump, hose and chemical) rig. It sported the best trim the decorators could muster. Favorite western colors of the time were white, battleship grey and Detroit grey. Detroit, Michigan, never had any grey rigs. The color name was a hold over from the New England Color and Paint Company in Detroit, which Seagrave owned, before the move to Columbus, Ohio.

The sketch of a Seagrave Gorham, pumping next to a vintage steamer, was done by Walt Pittman. Walt, a life-long Los Angeles area resident, has been a fireman, fire buff, historian, and Seagrave aficionado for fifty years. Letters from Walt often come with sketches like this one on the back. It seemed appropriate to include this grey tone sketch to remind us to note, record and photograph items in our lives, and our community's history, so that future generation's views will not have to be grey and sketchy.

These photos of W. E. Seagrave deliveries to Walkerville (Windsor) and Montreal depict the differences between the models R and Q. The R was a 144 hp, six-cylinder, and the Q was a 96 hp, four cylinder. Both were 7 3/4 inch bore, and 9 inch stroke. The R was a 900/1,100 gpm pump of three or four stages. The Q, rated at second size, 600/700 gpm, was a two stage pump.

Serial #10,000 was a milestone. The numbers on the sign boards represent purchase order numbers, however, and not apparatus. Order 10,001, for example, could have been for a spare hub cap. Stockton, California, received this 1914 delivery.

Three Gorham style pumpers were shipped to Scranton, Pennsylvania, in 1914. They were the 37th, 38th and 39th Gorhams.

ENGINE COMPANY No.1 S.F.D.
THE SEAGRAVE COMPANY
BUILDERS
COLUMBUS OHIO. U.S.A.
No. 10000

BUREAU OF FIRE
SCRANTON PA.
THE SEAGRAVE COMPANY
BUILDERS
COLUMBUS OHIO. U.S.A.
No. 12046

Here's another set of photos showing how rigs change while in service. Serial #12,112 was delivered to Chicago in the closing days of 1914. After a few years of service, the Chicago shops converted the apparatus to a pump and hose combination.

The die was cast both in engine and pump style. Homer Havens, the engineer from Seagrave, was equally impressed with the water cooled engines that powered the Gorham pumpers. Soon, Seagrave would adopt many Gorham engine features. Single cast cylinders, very long stroke design, multiple spark plugs, and a pump to engine pressure relief system, would all show up soon as standard Seagrave equipment.

There were a total of forty-one Seagrave Gorham pumpers built. The first sixteen were shipped to California for installation of pumps and engines. The remainder were assembled in Columbus. A beautifully restored, 1913 model R for Sacramento, California, is on display in the auto collection of the Imperial Palace, Las Vegas, Nevada.

The Gable Hood Era

1913 - 1926

The new style Seagrave at the Kansas City, Missouri exposition, 1919

The sign boards on the front of this sales demonstrator made a very honest comment. By the end of this era the competitor's gear type pumps would prove very ineffective. The efficiency and safety built into centrifugal pumps would dominate the fire equipment field.

Production in Columbus rose steadily during the gable hood era. Julius Stone was the major stockholder and chairman of the board. He had a strong work ethic from boyhood experiences on a farm in Michigan, mechanical knowledge from working on the railroad, financial success from other businesses, and recent success in Seagrave made Stone the right man to move Seagrave forward.

In San Francisco, the city's new model ladder truck negotiated steep hills at 10 to 12 mph, and 42 mph on the flat. Speeds of 60 mph were attained on an ocean beach run. Chief C. W. Rinker of Minneapolis set a department speed limit of 35 mph after placing two air cooled and one water cooled squad wagons in service.

The cost of keeping horses vs motorized fire equipment was in every trade journal of the day. Not only did it cost less to maintain, but with the addition of a greater response area in the same time frame, motorized equipment could eliminate the need for some fire stations entirely.

Springfield, Massachusetts is credited with being the first department to become fully motorized. Springfield and other cities kept careful records of fire equipment costs.

DAILY COST OF MOTOR-DRAWN APPARATUS	
Decatur, Ill	22 cents per day
Springfield, Mass	14 cents per day
Richmond, Va.	13 cents per day
Average daily cost	16 cents per day
DAILY COST OF HORSE-DRAWN APPARATUS	
Springfield, Mass	$1.90 per day
Richmond, Va	2.18 per day
Average daily cost	2.04 per day
Average gain per day of Motor Apparatus	$1.88

These 1914 photos show the busy shops. A wide diversification of product, and an ever increasing demand for the product led to an increase in factory square footage in 1910 and 1925 (see page 205).

The old buggy factory was humming like never before. The original structure contained 36,016 square feet. In 1910, coinciding with the closing of the Lane Avenue plant, an additional 18,096 square feet of office and floor space was added to this location.

The majority of the apparatus was Seagrave built. Only a few components were purchased, like radiators, from The Fedders Company. Seagrave always formed their own bodies and fenders, and assembled everything at 2000 South High Street, Columbus, Ohio.

In 1914, Seagrave produced ninety-three pieces of fire equipment. Only seven of these were localized Ohio deliveries. The quality of the product produced sales all over the country. Auburn, Maine; Bovey, Minnesota; Seattle, Washington; Phoenix, Arizona; Ada, Oklahoma and Augusta, Georgia demonstrate the wide delivery area.

The first water cooled gable hood Seagrave

Deliveries from the plant on South High Street were mixed and varied, but the apparatus in the photo, above, set the trend that would continue for years. This rig featured a water cooled engine ahead of cab configuration.

This hose and chemical for Astabula, Ohio from 1911, is a fine example of motorized equipment of the period. Apparatus serial #5858 came with a water cooled, six-cylinder D motor. The magneto was a Bosch, carburetor a Rayfield, and the engine featured a Hancock oil system. Firestone and Timken supplied axle components, and Rushmore supplied the lights. Seagrave designed and manufactured a vehicle around the best standard components and the best materials of the day. The body on this rig, for example, is made of aluminum coated steel. Delivered to Astabula, the cost was $7,500.

If you think that you have seen this photo before, you may be right. The Society for the Preservation and Appreciation of Antique Motor Fire Apparatus in America (SPAAMFAA) has been using this photo in its flyers for years. If you want more information about fire apparatus, or names of people who share an interest in old fire apparatus, write them at: P. O. Box 2005, Syracuse, New York, 13220.

In 1913, New York City, hosted the Fire Engineer's Convention. This 41st convention drew world wide recognition. New York, for example, had only one motorized piece of pumping equipment at the time.

Seagrave sent a chemical car and a ladder truck for static display and two pumping engines. The large Gorham, serial #A301, was to compete and the small 96 hp unit, shown at the right, was to give individual demonstrations. Other activities included seminars and a parade.

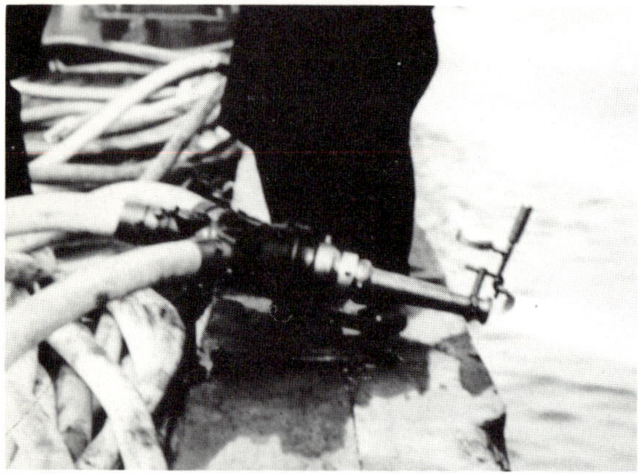

International Association of Fire Engineers. Convention: New York, September, 1913.

TESTS OF AUTOMOBILE PUMPING ENGINES, HUDSON RIVER, SEPTEMBER 3rd, 1913.

Under the direction of the Committee on Tests of the New York General Convention Committee with the co-operation of the National Board of Fire Underwriters.

The rated capacities of the engines submitted for test had been determined by the Tests Committee at preliminary trials during the previous week. The tests commenced at 6.30 a.m. and continued until 6.30 p.m. with only such intermissions as were needed for changing hose, nozzles, etc., but during such intermissions the motor was kept running and where possible the pump also. The engines were given the length of hose and sizes of nozzles as were found needful at the Committee's preliminary trial to ensure each engine maintaining the required pressure while delivering their rated capacities. Only one suction hosepipe was permitted with each engine. Engines rated at less than 700 A.G. (586 E.G.) used a suction hose not more than $4\frac{1}{2}''$ inside dia.; any engine rated at 850 A.G. (709 E.G.) used a 5″ section; rated at 900 A.G. (750 E.G.) or more used a 6″ suction hose. Not more than two men at one time were allowed to attend an engine, but the attendants could be relieved every three hours. Petrol of the same grade was supplied by the Committee for the use of all engines, and the amount consumed by each engine during tests was most carefully noted. Competitors were allowed to use any grade of lubricating oil they desired, the amount used, however, was not booked, although it was the intention this should be done.

SUMMARY OF TESTS. *(Compiled by our Special Correspondent.)*

Name of Engine / Type of Pump	Seagrave TURBINE	Ahrens Fox PISTON	La France ROTARY	La France ROTARY	Luietwieler PISTON	Knox PISTON	Robinson PISTON	Robinson PISTON	Nott ROTARY	Nott ROTARY	Waterous PISTON
Engine H.P.	144	79	72	126	57	60	94	94	68	101	101
Weight (in cwt.)	146	129	80	114	91	125	103	100	110	128	123
Wheel base	14′ 4″	12′ 4″	13′ 2″	14′ 6″	11′ 7″	14′ 4″	14′ 4″	14′ 5″	11′ 8″	11′ 8″	11′ 4″
Wheel track	5′ 8″	5′ 6″	5′ 2″	5′ 3″	4′ 10″	5′ 10″	5′ $0\frac{1}{2}$″	5′ 0″	5′ 11″	5′ 11″	5′ 9″
Time pumping in min.	720	720	615	652	454	643	687	540	537	691	720
Percentage of full time	100	100	85	90	63	89	95	75	74	96	100
Petrol used per hour A.G.	$18\frac{1}{4}$	$9\frac{3}{4}$	$10\frac{1}{2}$	$23\frac{1}{2}$	$7\frac{1}{2}$	$9\frac{1}{2}$	$12\frac{1}{2}$	11	12	14	$12\frac{1}{3}$
,, ,, E.G.	$15\frac{1}{4}$	$8\frac{1}{2}$	$8\frac{3}{4}$	$19\frac{1}{2}$	$6\frac{1}{4}$	8	$10\frac{1}{2}$	9	10	$11\frac{1}{2}$	$10\frac{1}{4}$

FIRST TEST.

Six hours pumping through three lengths of hose siamesed (direct) into one nozzle, at not less than 120 lbs. pump pressure.

	Seagrave	Ahrens Fox	La France	La France	Luietwieler	Knox	Robinson	Robinson	Nott	Nott	Waterous
Mean pump pressure	125	129	125	131	121	95	125	119	130	131	124
Mean nozzle pressure	48	63	61	87	42	42	55	53	64	49	56
Discharge per min. A.G.	1049	722	710	1405	508	508	780	865	631	832	680
,, ,, E.G.	816	575	570	1171	403	403	615	673	526	658	567
Loss (per cent.) in pressure from friction in hose	61	51	51	33	65	56	55	55	50	63	56
Size of nozzle	$2\frac{1}{4}$	$1\frac{3}{4}$	$1\frac{3}{4}$	$2\frac{1}{4}$	$1\frac{5}{8}$	$1\frac{5}{8}$	2	2	$1\frac{5}{8}$	2	$1\frac{3}{4}$
Water horse power	74	54	52	111	35	28	54	58	49	62	51
Efficiency	51	68	72	88	61	46	57	61	72	61	50

SECOND TEST.

Three hours pumping through a single line of hose, at not less than 200 lbs. pump pressure.

	Seagrave	Ahrens Fox	La France	La France	Luietwieler	Knox	Robinson	Robinson	Nott	Nott	Waterous
Mean pump pressure	204	210	220	209	195	232	214	213	201	185	218
,, nozzle ,,	80	61	90	64	45	25	55	64	44	49	90
Discharge per min. A.G.	596	360	355	727	309	392	415	448	306	392	355
,, ,, E.G.	470	285	280	582	244	310	327	355	240	327	296
Size of nozzle	$1\frac{1}{2}$	$1\frac{1}{4}$	$1\frac{1}{8}$	$1\frac{3}{4}$	$1\frac{1}{4}$	$1\frac{5}{8}$	$1\frac{3}{8}$	$1\frac{3}{8}$	$1\frac{1}{4}$	$1\frac{3}{8}$	$1\frac{1}{8}$
Length of hose in feet	150	500	400	100	700	600	400	300	700	400	400
Loss in pressure from friction	61	71	59	69	77	89	74	70	78	74	59
W. H. P.	69	43	44	87	34	51	50	54	34	43	46
Efficiency	48	54	61	69	60	85	53	57	50	43	46

THIRD TEST.

Three hours pumping through a single line of hose, at not less than 250 lbs. pump pressure.

	Seagrave	Ahrens Fox	La France	La France	Luietwieler	Knox	Robinson	Robinson	Nott	Nott	Waterous
Mean pump pressure	256	260	267	268	Withdrawn	262	225	Withdrawn	Withdrawn	249	259
,, nozzle ,,	74	85	117	64		29	100			45	109
Discharge per min. A.G.	323	345	321	626		359	295			251	307
,, ,, E.G.	254	272	252	502		283	233			200	243
Size of nozzle	$1\frac{1}{8}$	$1\frac{1}{8}$	1	$1\frac{5}{8}$		$1\frac{1}{2}$	1			$1\frac{1}{8}$	1
Length of hose in feet	700	600	600	200		700	600			1300	600
Loss in pressure from friction	71	67	56	76		89	56			82	57
W. H. P.	46	50	48	96		53	37			33	45
Efficiency	32	63	67	76		88	39			33	45

Wednesday, the third day of the convention, was perfect for throwing water. It was cloudless and windless. It was so warm, that buckets with ice were provided to refresh the engineers. The wharf at the end of 54th Street was made available for testing . A huge score card was assembled, and all those in attendance were given information sheets on how the tests were to be conducted.

The engines were positioned on the pier with five on one side and six on the other. Because of the tide, maximum lift was 14 ft. 4 in. at 10:30 a.m., and 7 ft. at 4:50 p.m. The delivery hose was laid out and the nozzle clamped to the dock. Pressure gauges were fitted to the nozzle. During the test, readings were taken every five minutes. A steamer was placed in the middle of the dock, and it's whistle was the signal to take the readings. Conditions for the test were from the National Board of Fire Underwriters.

The apparatus had to be able to deliver it's full rated capacity at 120 pounds pressure and half it's capacity at 200 pounds pressure. Hose lengths varied according to the test (note the chart at left). After testing, engines and pumps were disassembled, and the conference attendees were allowed to inspect and compare for wear. The Seagrave engine performed flawlessly. It pumped continuously for twelve hours, consuming 219 gallons of gas in the process.

The Luietwieler, Robinson, and 68 hp Nott could not complete all of the tests. Other apparatus suffered minor problems. Only the Ahrens-Fox, Seagrave, and Waterous were rated 100 percent reliable. In overall marks, the Seagrave placed behind some other makes based on fuel economy. The Ahrens-Fox used 117 gallons of gas, and the Waterous used 148 gallons.

The centrifugal pump had met the challenge, and only needed to be slightly refined to be the overall most efficient.

Seagrave engineers were quick to learn from their New York City test results. Engine efficiency must increase and hose carrying capacity must be enlarged. This would mean that the huge Gorham engine would have to be downsized or modified.

Increasing hose box size was accomplished by mounting a Gorham two stage pump lower in the frame. As indicated in these deliveries for Chicago, Illinois, above, and San Diego, California, left.

The new Chicago rig, on its acceptance test, delivered 473 gpm through a stand pipe to the top of the twenty-two story Masonic Temple (22 stories = 356 feet). The Chicago pumper was the 54th centrifugal, and Seagrave advertised "Not a single repair part has been ordered for a Seagrave centrifugal pump".

The rotary gear pump of American LaFrance and the piston pump of Ahrens-Fox were carry overs from steamers. The designs were well refined from years of service, and would be worthy of inspection. But, Seagrave would stay with the centrifugal.

Further testing at Seagrave, in 1914, included this unusual pumper for Saginaw, Michigan. Or, was it Fairfield, New Jersey? The question arises from the odd note etched into the negative. This unique apparatus has double row louvres on the hood, and Ross relief valves on the discharge gates, not typically found on Seagrave equipment. The rest of the apparatus appears to be Seagrave and it can only be assumed that his was an experimental model.

The Gorham pump was to be replaced by a new pump, the Rees Rotorbo. The Manistee Iron Works of Manistee, Michigan, was making centrifugal pumps under license from Rees, an English concern. The basic shape and size could be changed to accommodate 2, 3, or 4 stage pumps, and varying pump volumes.

The standard Manistee pump, right, was modified by Seagrave to become a hollow shaft drive pump. Power to the pump was through a tube. Power to the wheels was through a solid shaft running through the hollow tube. The cut-away photo, on page 109, will explain the hollow shaft drive principle.

FIG. 2—REES RoTURBo FIRE PUMP FOR CITY FIRE ENGINE.

Seagrave exhibited at two major shows in 1915. The Fire Engineer's Convention in Cincinnati, Ohio, and the Panama-California Exposition in San Diego, California. Seagrave excelled at both.

The San Diego exhibition featured the new Seagrave engines. Based loosely on the Gorham design, they featured cast iron single cylinders on a cast brass crank case. They were referred to as "T" heads because of the basic shape of the cylinder casting or jug. The first applications of the new engines were the special rigs shown for the exposition, shown at the top of the page.

At Cincinnati, three pumping engines were displayed: A) a 1,000 gpm, four stage pump with 11 inch impellers and the new H-6 engine that had a 6 3/4 inch bore, and 8 inch stroke; B) an 800 gpm, four stage pump with 10 inch impellers and the new F-6 engine of 5 3/4 inch bore and 6 1/2 inch stroke; C) a 500 gpm, four stage pump with 9 1/2 inch impellers and the F-4 engine.

These pumpers featured the new hollow pump shaft "Manistees". The lineup from late 1914 through February 1920, with few exceptions, would be:

Engine model	Pump Size (gpm)
F-4	325-500
F-6	500-750-900
H-6	1,000-1,200

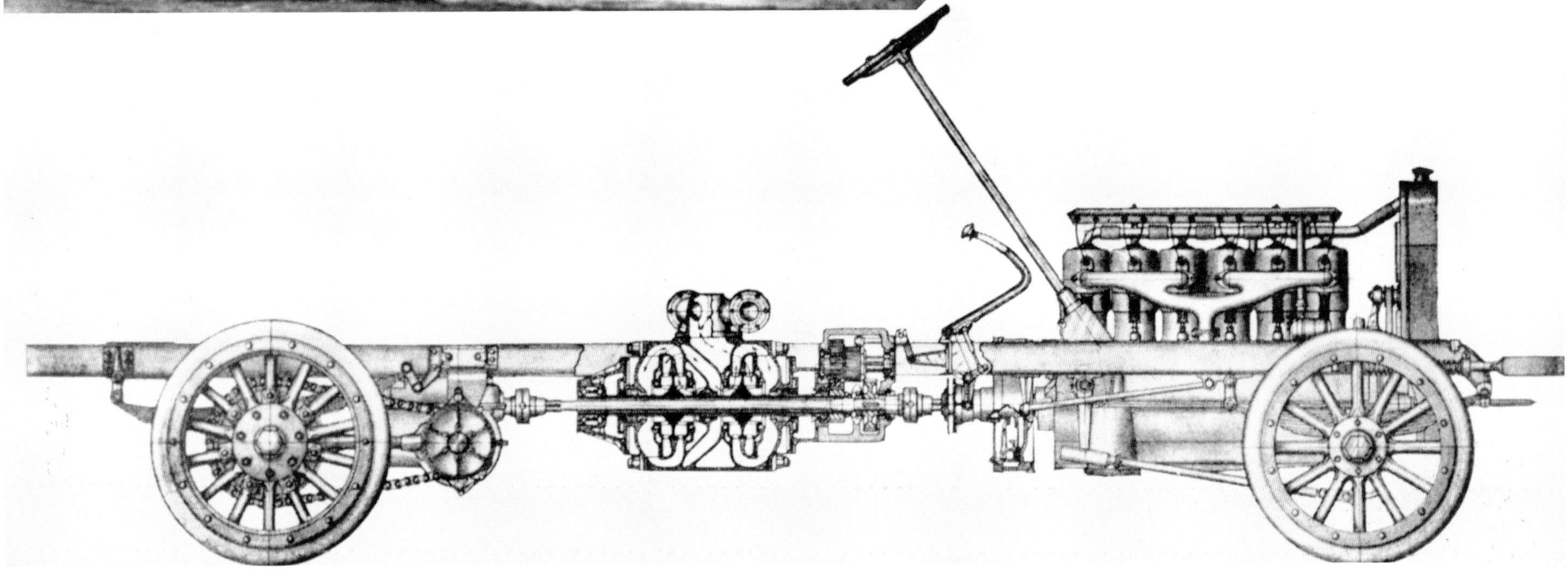

The Gorham style pumper, shown at a pump test at the Detroit River, would soon give way to the Manistee style pump, shown in the cross section. Through the late teens, this would be the standard layout; a midship mounted, hollow shaft pump. The engagement lever for the pump was under the driver's seat. With slight changes in components, this basic layout would survive for more than 50 years.

Charleston, South Carolina received this 1,000 gpm pumper in August, 1915. Stated on the work order was "must pump 1,000 gpm at 120# - 600 gpm at 200# - 500 gpm at 250#".

This could be accomplished with the larger type H motor. From these factory photos, the way to tell a type H motor is that the top of the headlight is below the break line in the hood, as opposed to the F motor for Framingham, where the hood line is below the top of the headlight.

Charleston's rig featured a "Sireno" brand electric siren, an oil sight gage on the dash, and a large, 20" diameter steering wheel.

Framingham, Massachusetts received this 750 gpm pumper. This apparatus was on a 172 inch wheelbase, with an overall length of 21 ft. 8 in. The widest part of these rigs were the rear hub caps, which stuck out four inches on each side of the 60 inch rim-to-rim dimension.

Chicago received two pumpers from Seagrave, in 1915. This rig was assigned to Engine Company No. 2, Department No. 32. It remained in service until 1934, when it was destroyed at the Chicago stock yard fire.

Houston, Texas, received this pumper in mid 1916. It displays the next pump improvement. Dual discharges direct from the pump increased water flow. This design casting was in effect until 1920.

By 1915, the product mix, from Columbus, was half motorized pumpers. But, cities, large and small, continued to order chemicals, ladders, and hose cars. These non-pumping rigs used the F2000 road transmission, which raised the jack shafts higher and closer to the frame rail.

Each apparatus had its own powertrain layout. Compare the jack shaft locations on these chemical for Malden, Massachusetts and Toledo, Ohio, with pumping engines on the previous page. Close examination of the photos show refinements from the bumpers to the drive sprockets. Seagrave included changes as they occurred and did not wait for a new model year.

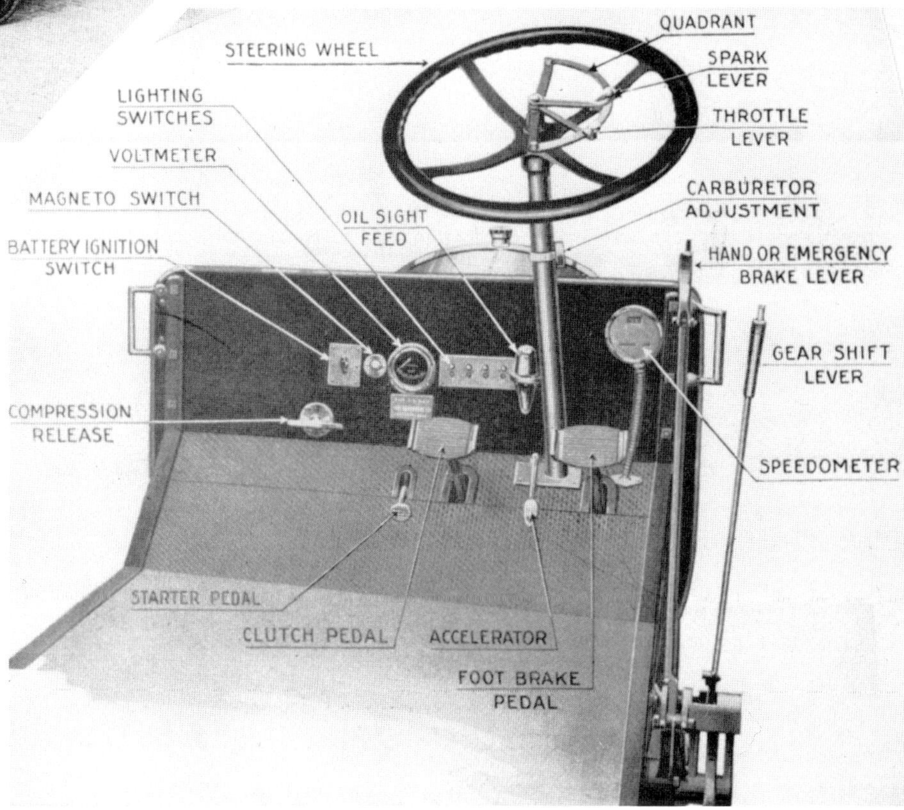

The Seattle, Washington Fire Department, in 1916, boasted of the following Seagrave equipment: (L to R) A) Hose 31, serial #14581, rebuilt in 1930, retired in 1955; B) Hose 10, serial #14578; C) Squad 1, serial #14584, delivered white, later repainted red, was a demonstrator at the 1915 San Diego Convention; D) Hose 2, serial #14579, served until 1955; E) Hose 5, serial #14580, also sold in 1955 - a very enviable length of service.

Although driving looked simple, lack of power steering, power brakes, and synchromesh transmissions made even the shortest trip a real workout. Seagrave's most common power plant was the F-6, rated at 79.3 hp. This horsepower rating is deceiving because the figure was arrived at by a strictly mathematical formula, and not measured at the flywheel. The formula was:

$$\frac{\text{diameter of cylinder}^2 \times \text{number of cylinders}}{2.5}$$

True flywheel horsepower of this engine was 116. The dual system was in place for years because some states classified and taxed trucks by horsepower.

Centrifugal Pump and Pump Transmission Assembly.

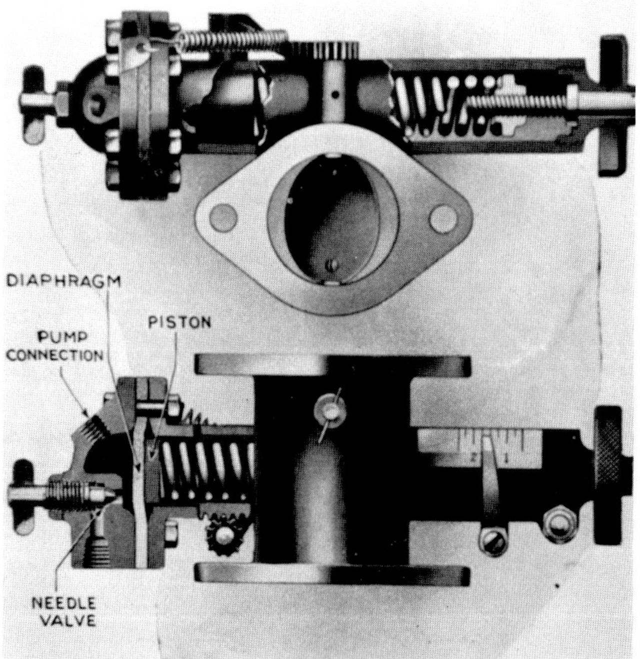

The Seagrave Pressure Regulator

OFFICIAL

International Association of Fire Engineers

44th Annual Convention

Providence, R. I., August, 1916

Test of Automobile Pumping Engines at State Pier, August 30, 1916

Under the direction of the Committee on Exhibits of the Association, with the assistance of Engineers of the National Board of Fire Underwriters and of the New England Exchange.

("All pumps drafted water with suction lift from 8 feet, 9 inches to 15 feet, 6 inches")

Pumper No.	NAME OF MANUFACTURER	MOTOR				TYPE OF PUMP	Rated Capacity at 120 lbs. Net Pump Pressure	Net Pump Pressure Attained	Gallons Discharged	Percentage of Rated Capacity Discharged
		No. of Cylinders	Bore	Stroke	H. P. A. L. A. M.					
Test No. 1—at 120 Pounds. Duration of Test—6 Hours										
1	American La France	6	5½	6	72.6	Rotary Gear	1000	131	1010	101%
5	Ahrens-Fox	6	5½	6	72.6	Piston	900	132	931	103%
4	**SEAGRAVE**	**6**	**5¼**	**6½**	**79.3**	**Centrifugal**	**750**	**138**	**788**	**105%**
Test No. 2—at 200 Pounds. Duration of Test—3 Hours										
1	American La France	6	5½	6	72.6	Rotary Gear	1000	221	516	51.6%
5	Ahrens-Fox	6	5½	6	72.6	Piston	900	216	516	56.5%
4	**SEAGRAVE**	**6**	**5¾**	**6½**	**79.3**	**Centrifugal**	**750**	**211**	***534***	***71.2%***
Test No. 3—at 250 Pounds. Duration of Test—3 Hours										
1	American La France	6	5½	6	72.6	Rotary Gear	1000	261	423	42.3%
5	Ahrens-Fox	6	5½	6	72.6	Piston	900	271	411	45.6%
4	**SEAGRAVE**	**6**	**5¼**	**6½**	**79.3**	**Centrifugal**	**750**	**258**	**355**	**47.3%**

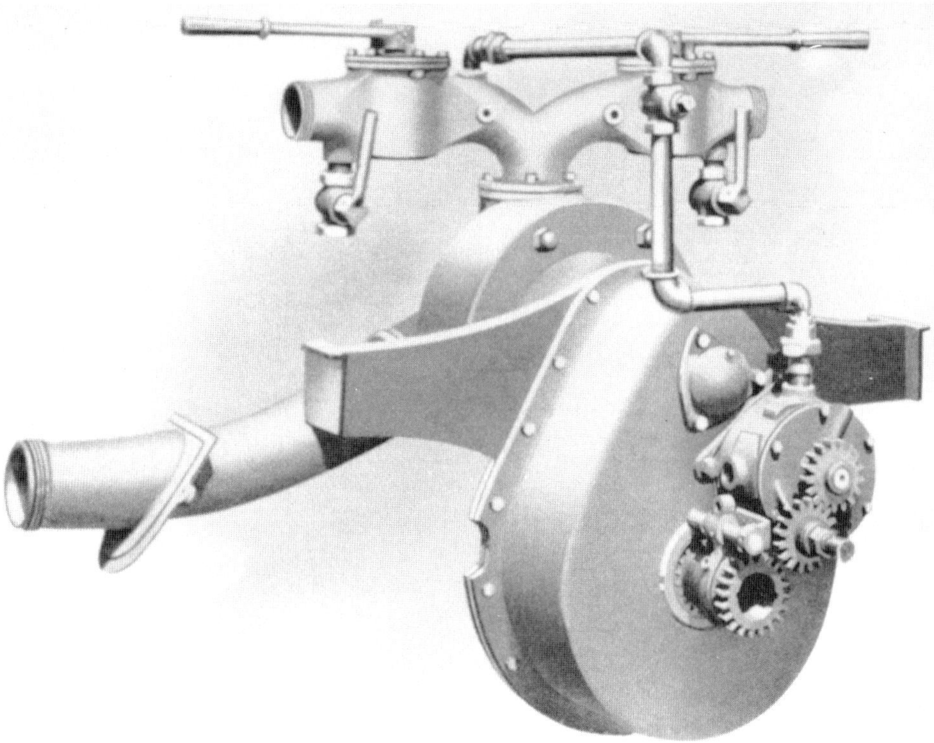

Fig. 66. Pump Assembly.

For the convention in 1916, Seagrave used serial #15060. Shown left, are two unique Seagrave features, the pump transmission with built in primer pump (first large gear to the right of the input shaft), and the automatic pressure regulator. The regulator automatically throttles down the engine if a discharge line is shut off, thus maintaining constant line pressure to remaining lines.

Photos, to the right, show the small 325 gpm pump. The single stage pump was a favorite of many small towns. Wauseon, Ohio's rig featured this small pump with a 3 inch intake and a 15 foot suction hose. Shown here, at a recent muster, it has been repainted, but is still in its original auto-grey color.

Serial #23310 was an F-4 powered apparatus for the 1919 Kansas City Convention. Easton, Pennsylvania received their F-6 powered apparatus in February, 1916.

The overall length of an F-4 unit was 19 feet. The F-6 was 20 feet 4 inches long. The size variation can be noted by comparing the length of the hoods to the length of the fenders.

Custom builders allowed the customers much freedom. Some customers ordered the oval gas tank (capacity 20 gallons, see page 115), some ordered the round tank (capacity 30 gallons, see page 114), and some ordered the tank shown here (40 gallon capacity).

Painting and striping was to the customer's order. The most common color was coach painter's red, with white or grey a distant second. Some communities, like Chicago, used only initials on the hood, and some cities, like Allentown, Pennsylvania, used initials on the hood and the name of the fire company on the hose box.

C.F.D.

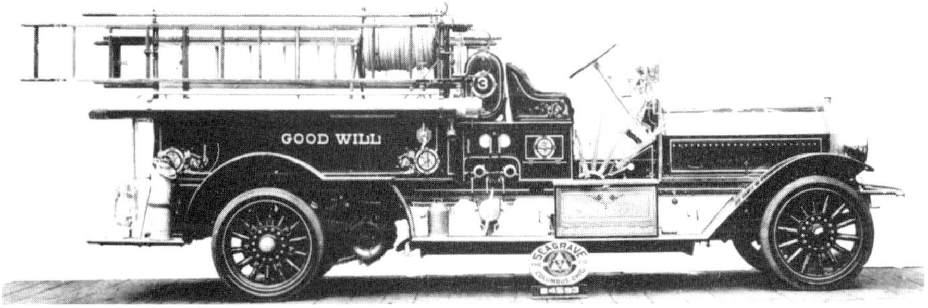
GOOD WILL
SEAGRAVE

LIBERTY
SEAGRAVE

COLUMBIA
SEAGRAVE

FRANKLIN
SEAGRAVE

In February, 1920, Seagrave started to produce a new style Manistee pump. Designated the “Manistee B”, this B style pump would be the mainstay of Seagrave for the next fifty years. The first delivery of the new style pump went to Santa Barbara, California. The top photo view shows the new style Manistee pump layout.

This new pump was more efficient. It allowed the F-6 engine to now power a 1,000 gpm pumper. The F-6 engine and a pump system rated at 750 gpm or 1,000 gpm would soon account for the lions share of the business. The new style pumpers were 100 pounds heavier and cost $500 more than previous versions. Weights and costs averaged 13,600 pounds and $11,500.

Centrifugal pumps made the engineers job easier. Long periods of pumping were less eventful than with piston or gear pumps.

The new Manistee solid shaft pump replaced the hollow shaft pump in 1920. This new style pump was mounted higher in the frame (compare location with page 109). The new, more efficient, pump allowed the F-6 engine to pump 1,000 gpm. The large H engine was relegated to the new 1,300 gpm Metropolite, which debuted in 1921.

With the new style, in 1920, came other noticeable changes. The pump control panel, with gauges and discharge gates, was also moved to just below the operator's seat.

The new road transmission was mounted higher in the frame. Gone was the "over and under" transmission that allowed you to tell the difference between a pumper and a hose wagon at a distance. The new transmission's jack shafts were fitted with 13 1/ 2 inch brake drums, thus making it a little easier to stop this 13,000 pound apparatus.

The sketch, shows how water is flung by centrifugal force to the outside of the pump. The flow is then directed to the center of the next impeller, and the water's movement is enhanced again. The void created by the vacating water is filled by water coming in from the original source.

The centrifugal pump is not self priming. This means its initial supply must come from a hydrant, or a small, positive displacement pump used only for the initial prime. Seagrave accomplished this with a small vane type pump piped into the main pump housing.

The photo, at the right, shows a cut-away training aid. Near the center of the photo is the shift lever. The handle to the right of the lever is rotated clockwise almost 180° to lock the clutch in place and prevent slippage while pumping.

The handle to the left of the lever sets the driveline and pump for engagement. The three slots for the handle are road, prime or pump.

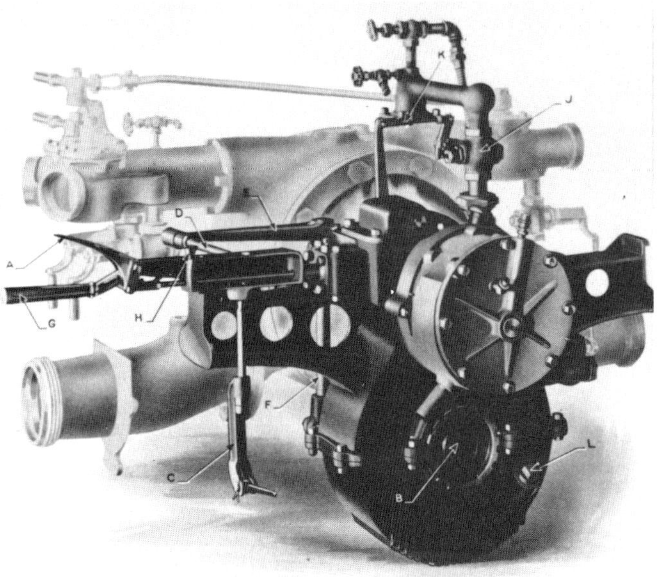

Seagrave pump transmission

The Seagrave primer pump and pump transmission are one unit. The primer pump, in the foreground was in use only when the pump handle initiated its use. It was a small vane type pump.

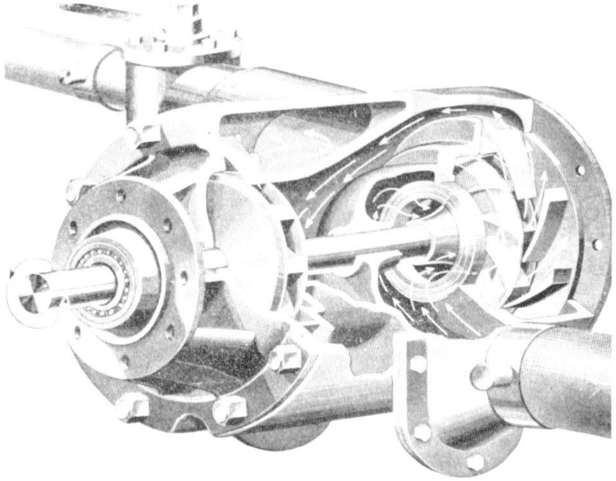

Our centrifugal pumps are guaranteed to deliver their rated capacity for a period of ten years - - -

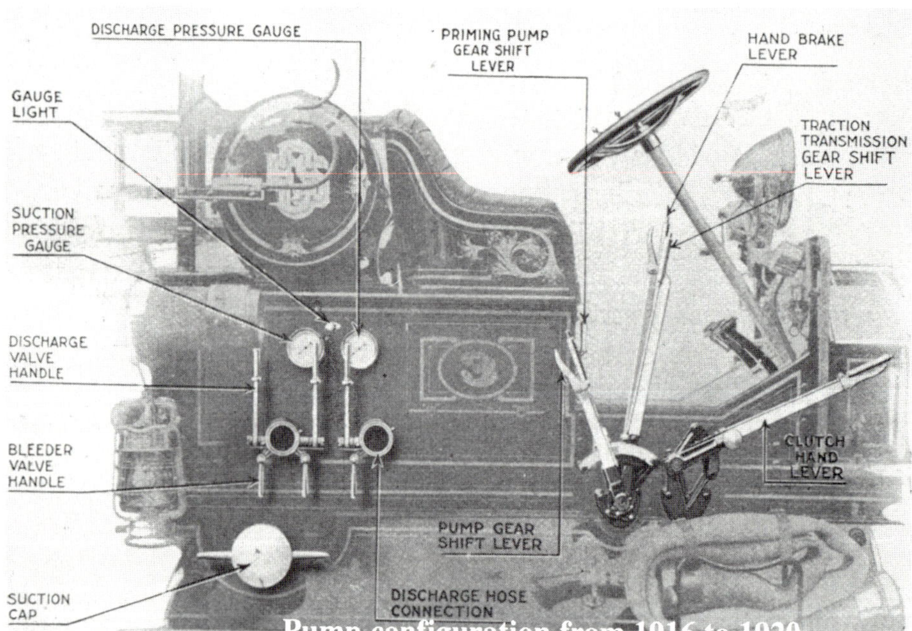

Pump configuration from 1916 to 1920

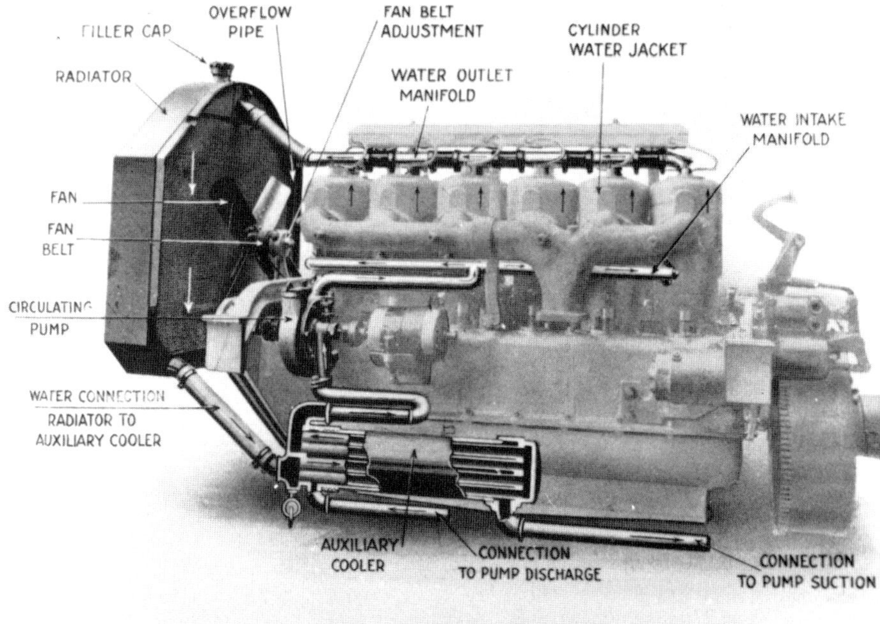

Tokyo, Japan, ordered fourteen pieces of apparatus, in 1919, four ladders and ten pumping engines. The pumpers were the 325 gpm F-4 powered units, shown top left. The F-4 engine was simply an F-6 with two cylinders removed. The engine rated 52.8 hp and true brake horsepower was 77.

The U. S. Army delivery, in 1919, and the Philadelphia, Pennsylvania, delivery, in 1920, help define the changes in pump location.

The F series engine went through many design changes, shown here are the early 1920's versions. The F series would adopt, in 1921, a pipe that extended gasses to preheat the intake manifold. Early gasoline quality was poor and the warm intake manifold helped vaporize the gasoline. The valve lifters became roller lifters and there was the option of three spark plugs per cylinder, one battery - two magneto fired.

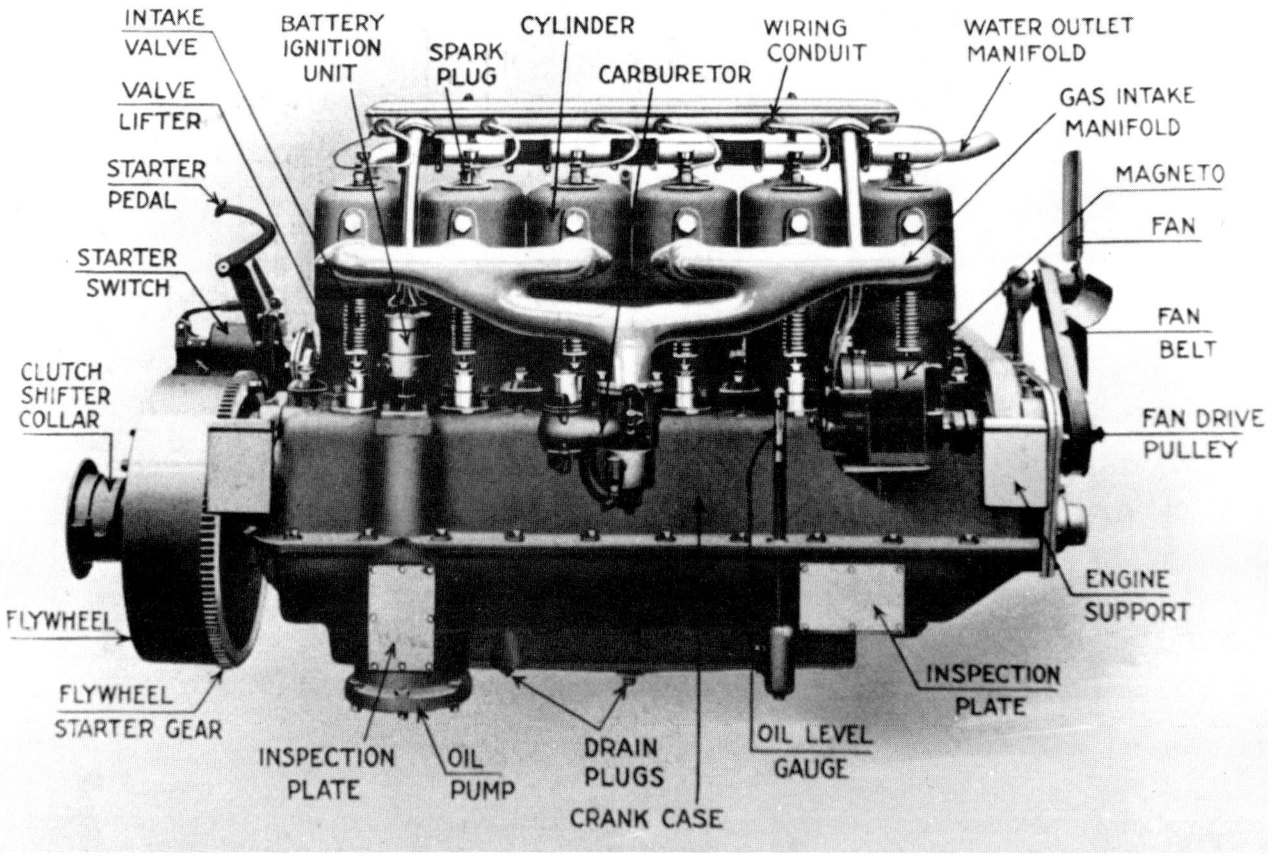

Huntington Park, California's 1921 Seagrave dominates the scene in this vintage shot. The pride of the department is flanked by a 1928 Buick and a 1922 Reo Speedwagon pump and hose car. Most departments learned "anything can haul hose, but when it comes to pumping water, get a Seagrave."

Another mystery rig. The work orders from the factory are missing, but the painter's log indicates that serial #26098 went to Tokyo, Japan, in October, 1920. A little detective work revealed that politics intervened. The Tokyo order was cancelled, and the apparatus was refitted to the specifications of Wellsburg, West Virginia.

Montreal received Model F pumpers, serial #28251 and #28252. The photo shows the start of a twelve hour acceptance test, June 17, 1921. An acceptance test on these powerful machines was almost boring. The very nature of the centrifugal pump made it quiet and well balanced. The typical American LaFrance of the day, with its gear hum, was noisy. The Ahrens-Fox, with its piston pump, would literally dance on the ground. The front end of the machine would sway to the point that many departments would strap wooden shoes or plates to their soft suction hoses, so that they wouldn't abrade from continuous movement on the pavement.

Subtle changes in style and mechanics can be noted starting in 1920. The switch to cycle style, (wheel hugging) front fenders, and the greater use of air tires vs the solid tires on pumpers. Because of their more stable nature, solid tires were recommended on aerial ladder trucks and water towers.

During the teens and early twenties, gongs were replaced by "bells and whistles", a term still used today to describe the extras on almost anything.

The bell was a locomotive style bell. Its ring could be heard above the traffic on the way to the fire. At the fire scene the bell could be used as a signal to evacuate the building. The sound of the bell could overcome the din of most fires. The whistle was a brass steamer type that ran on exhaust gases.

The hand cranked siren started showing up on apparatus, in 1910. The siren was usually operated by the officer sitting next to the driver. A hand cranked siren is shown, mounted to the dash, on this early gable hood apparatus. By thc late 1920's, electric sirens replaced the hand cranked versions.

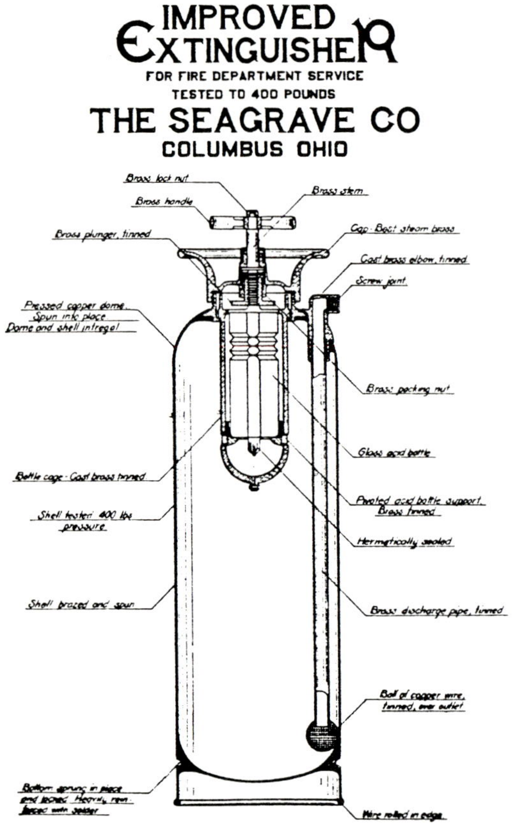

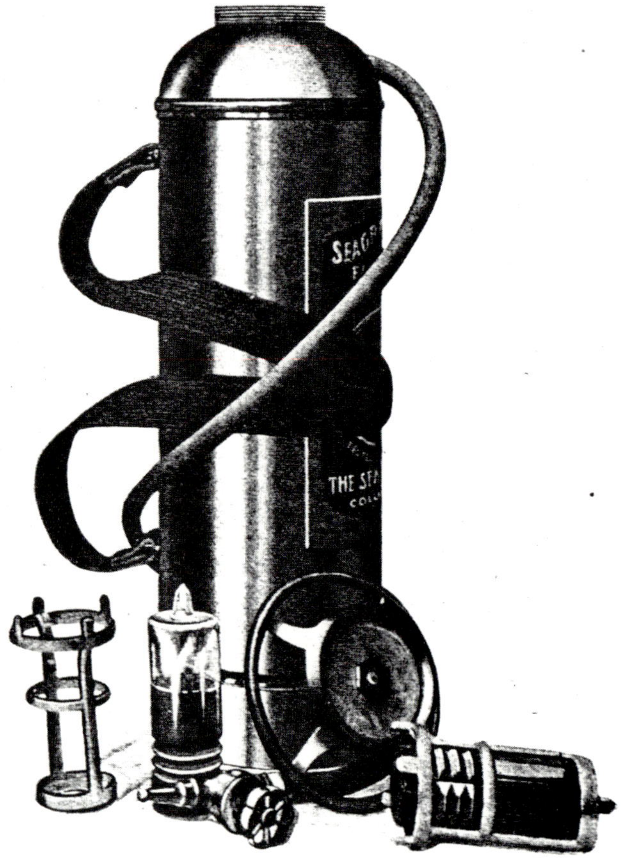

Large towns or small, a great percentage of fires were extinguished with the hand-held soda acid extinguisher. Seagrave made their own.

The apparatus extinguisher differed from the more common building extinguisher in three ways. First, a screw on the top of the extinguisher forced the lead stopper of the acid bottle to remain seated during transit. Second, a shut off on the discharge hose enabled the fireman to start and stop the stream. Lastly, a shoulder strap allowed for free use of both hands at the fire scene.

Shown on these pages are a combination hose and ladder truck and a pumping engine for Methuen, Massachusetts. Both carried Seagrave extinguishers. The long ladders on a short truck looks odd to us today, but in the early days, in many areas, this was the norm (see the following two pages).

The pumping engine for Methuen was delivery tested from a water cistern, a ten foot lift. This 750 gpm pumper delivered 788 gpm at 120 psi pressure with an engine speed of 1,030 rpm. On the high pressure test it delivered 367 gpm at 200 psi pressure. The ladder truck received with this order appears on page 141.

SEACO
TESTED
TO 350 POUNDS
TO CHARGE
TO OPERATE
THE SEAGRAVE CORP'N.
COLUMBUS, OHIO

SEAGRAVE
COLUMBUS, OHIO

The largest city in Pennsylvania, Philadelphia, and one of the smallest, Monessen, used Seagrave chemical cars. The chemical tanks were made of copper. The seams were riveted and soldered. The inside of the tank was then coated with lead.

The apparatus on these pages again highlight the chemical apparatus and its variations. Seagrave sold complete rigs, or would install Seagrave equipment on the chassis provided by the customer. A 1917 commercial car quote read:

2	20 gallon chemical tanks, hose, reel, and nozzles	$818.00
2	Lanterns with holders	15.00
2	6 lb. pick axes	10.00
2	3 gal. extinguishers	20.00
	Paint - F.D. Red	30.00
1	20 ft. extension ladder with brackets	60.00
	Total	$953.00

Seagrave never promoted commercial fire apparatus. When asked to build on a commercial, the quote was always "pricey". The Hudson car dealer near by Newark, Ohio, tried to spur Hudson sales with the touring car conversion, shown above.

Tracing the history of apparatus is often tricky, as was this delivery to Ford, Michigan. Ford is one of those towns that does not exist today. Many communities were absorbed by neighboring communities. This usually took place during times of economic crisis. The town of Ford didn't make it through the economic crisis of 1921. It is part of Wyandotte, Michigan today. Its old station house became a library. The rig itself was declared surplus and sold to Elkhart, Indiana.

Detroit, Michigan, has always been a big Seagrave user. Well over 200 pieces of motor apparatus have been purchased.

Over the years, Detroit was also home for the Sewell Cushion Wheel Co. The Sewell wheel contained a pliable and resilient rubber filler between a hard wheel and a hard outer rim. This afforded some shock absorption without the soft ride of air tires.

Serial #14474 was the last all chemical apparatus ordered by the city. Serial #14476 was assigned to Engine 1 and saw service in various locations, until 1948, a career of 33 years. Detroit's distinctive black oval on the hood always remained blank. It was never used as a background for a company number.

D.F.D. on the hood doesn't always mean Detroit, Michigan. Dayton, Ohio, and a host of cities used D.F.D. on the hood. Also, there are cities named Detroit in Alabama, Illinois, Kansas, Main, Oregon, Texas and a Detroit Lakes, Wisconsin.

Boston Fire Commissioner, Benjamin W. Wells, directed the modernization of the Boston Fire Department, starting in 1907. Before this time, Boston had two self propelled steam pumpers, but they were so slow they were only called out on second alarms. The first motorized apparatus came on converted passenger car chassis, a Knox and a Packard. Over time, Boston and most other cities made the switch to custom fire apparatus.

The upper photos show how the ravages of winter weather can cause extra hardship. Chemical engines 8 and 36 were both delivered, in 1917. They are shown in action at a Charlestown section of Boston fire, January 18, 1920. Close inspection of the photos show how ice would build up on the apparatus during winter blazes.

Often apparatus that was iced in had to be chopped free. Sometimes a tent like structure, complete with wood stove, would be built over the ice bound item. Thawing could take longer than the fire service time itself.

In 1916 and 1917, Boston received multiple orders from Seagrave, including serial #16326 hose and chemical #10. The photo, above, shows the apparatus as it appeared in 1922 in front of the station it shared with Ladder 4, at 194 Dudley Street in the Roxbury section of Boston.

The venerable Seagrave aerial ladder has been towed around by almost everything. Pictured top to bottom are: serial #9860, a Knox-Martin three wheel tractor for Pittsburgh, Pennsylvania; serial #24396 , a Knox tractor for Cleveland, Ohio; and a U.S.A. tractor for an early Seagrave horse drawn trailer in Covington, Kentucky.

Rochester, New York's, tractor looks very typical. Especially when compared to serial #10210, an electric drive for Grand Rapids, Michigan. Of all the methods shown on these two pages, the long wheel base Seagrave tractor provided the most stability and ease of operation. Properly braced, the tractor becomes the counter balance for the aerial ladder.

K model Seagrave tractors came in four and six-cylinder versions. Most were powered with the H series engines. The exception was five New York City deliveries with F-6 engines, shown lower right. The front drive tractor's salability came mainly from the fact that they could extend the useful life of a horse-drawn steamer, ladder, or water tower. The concept sold well, and some cities preferred to order the stubby tractor as part of the new aerial combination.

Seattle, Washington ordered two model K tractors in 1915, one for steamer conversion, the other for attachment to their water tower. Two views of the steamer tractor are shown. These tractors were powered by the H-4 motor of 6 3/4 inch bore and 8 inch stroke, that produced 110 brake horsepower. The Rayfield carburetor on these units had a 2 1/2 inch throat. Tires were 7 inches wide and mounted on 38 inch rims.

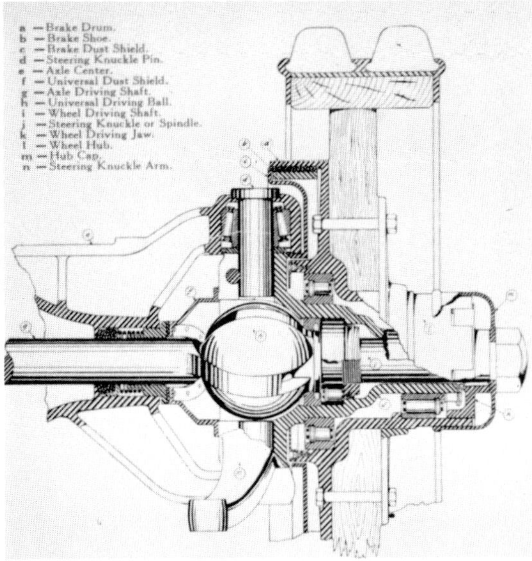

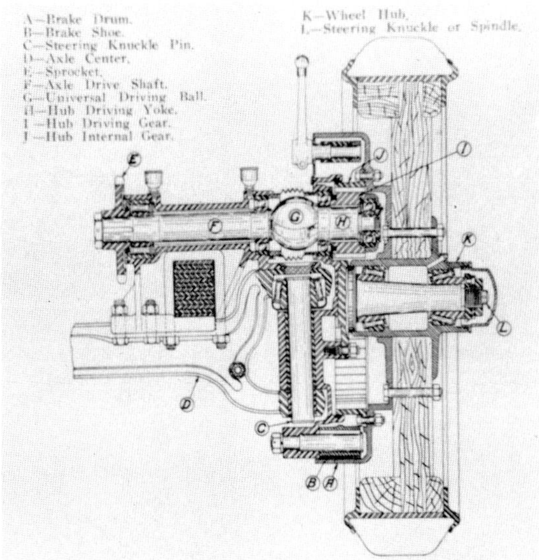

Fig. 74. Two-Wheel Steamer Tractor Axle and Wheel Hub.

There were two distinct types of steerable drive axles. The configuration in the top photo shows the arrangement for a heavy front drive application or a rear drive application.

The lower photo shows the gear reduction in the wheel type drive axle. This was the most common style.

Shown here are model K front wheel drives. Seagrave would provide just the front drive unit with frame extension or a complete apparatus.

The short hood version, for Seattle, Washington, was powered by an H-4 engine. The F or H six-cylinder engine required a hood extension.

Chambersburg, Pennsylvania, was proud of the new equipment at Junior Hose & Truck Company 2. Both the pieces of apparatus were 1917 Seagrave deliveries.

Interdepartmental rivalry is as old as fire department history itself. Chambersburg's first fire company, Friendship, was formed in 1780. In 1855, some boys that were rejected by the Friendship Company, as too young, formed the Junior Company. The friendly rivalry continued for many years. The rival Friendship Company used American LaFrance equipment.

Philadelphia, Pennsylvania, also used the K chassis for some of their ladder equipment, like Ladder 4, shown at the right.

Note the tiller wheel is already cocked. The driver turning left has already started to swing the frame left. The tiller driver has had to turn his wheels right to overcome the frame's left tendency. After clearing the door, the tiller driver would then have to swing hard left to bring the tail around.

The individual photos on these pages depict Seagrave equipment with the H series engines. Denver, Colorado, because of its altitude, always needed equipment with large displacement engines.

The power of a gasoline engine is reduced at high altitude. The air - fuel ratio through the same size carburetor openings, is not the same at 5,000 feet of altitude as it is at sea level.

The Denver, Colorado, City Hall is the backdrop for the main photo on this page. The main fire station is on the first floor of City Hall. Almost all of the fire equipment shown, in this 1920 photo, is Seagrave. Through the teens, Denver's department began to resemble a Seagrave catalog. They had buckboards, model K's and gable hood deliveries.

At Denver, as in so many of the cities, the records of all the older apparatus were kept until just a few years ago. Then, with storage space at a premium, the early records were destroyed.

Four Wheel Drive and Steer Aerial Truck

Straight Gasoline Type. Built in three sizes: 65, 75 and 85 feet.

City Service Truck — Style J

Two Wheel Front Drive Aerial Truck

Built in four sizes: 55, 65, 75 and 85 feet.

Four Wheel Tractor Drawn Aerial Truck

Built in four sizes: 55, 65, 75 and 85 feet.

City Service Truck — Style F

Four Wheel Tractor

By the teens, Seagrave product mix covered a wide spectrum. With styles becoming more complex a model code was necessary. The code was a series of letters and numbers:

The letters indicated:	The numbers indicated ladder length:
F - straight frame ladder	65 = 65 foot aerial
J - tiller straight frame ladder	65 = 75 foot aerial
K - front drive tractor	85 = 85 foot aerial
L - four wheel tractor	
M - side steer ladder	
Q - four-cylinder Gorham style pumper	or engine type:
R - six-cylinder Gorham style pumper	
S - H engine Manistee style pumper	4 = four-cylinder engine
T - F engine Manistee style pumper	6 = six-cylinder engine

Add the letters to ladder length or engine size. Example: a model L756, would have been a tractor-drawn 75 foot aerial ladder with a six-cylinder engine.

Combination Pumping Engine and Hose Car

Capacity, 300 to 375 gallons per minute.

Combination Chemical and Hose Car—6 Cylinder

Combination Pumping Engine and Hose Car

Capacity, 500 to 600 gallons per minute.

Two Wheel Front Drive Tractor

Heavy Steamer and Aerial Truck Type.

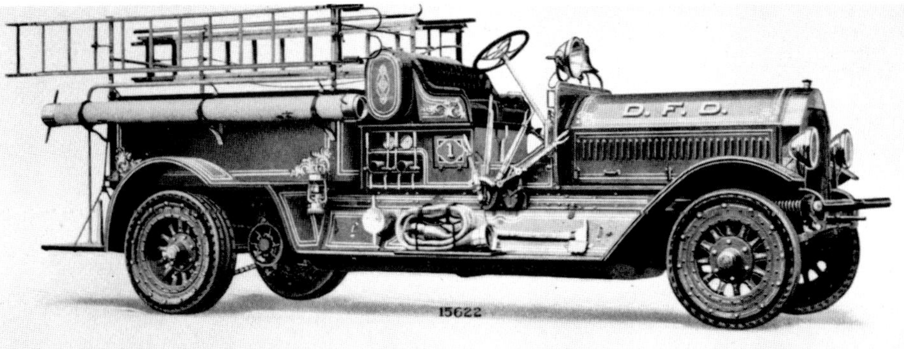

Combination Pumping Engine and Hose Car

Capacity, 1,000 to 1,200 gallons per minute.

Squad Car

The product mix of the early 1920's favored the tractor-drawn aerial and the straight frame service ladder truck. Prices ran: $6,000 for a tractor to put under an existing ladder, $9,000 for a straight frame service ladder, and $17,500 for an 85 foot tractor-drawn aerial ladder truck.

By this time Seagrave discouraged the purchase of the two wheel tractor or tiller rear steer on straight frame trucks. The tiller on a straight frame ladder required the steerable drive axle (see page 133). It was an expensive $1,300 option, and it posed driveability problems. The tractor trailer was preferred because the tractor could be used as a counter balance weight for the aerial.

Shown are deliveries for: Attleboro, Massachusetts; St. Paul, Minnesota; Methuen, Massachusetts; Petersburg, West Virginia and Bryn Mawr, Pennsylvania.

LADDER
METHUEN
SEAGRAVE
COLUMBUS, OHIO

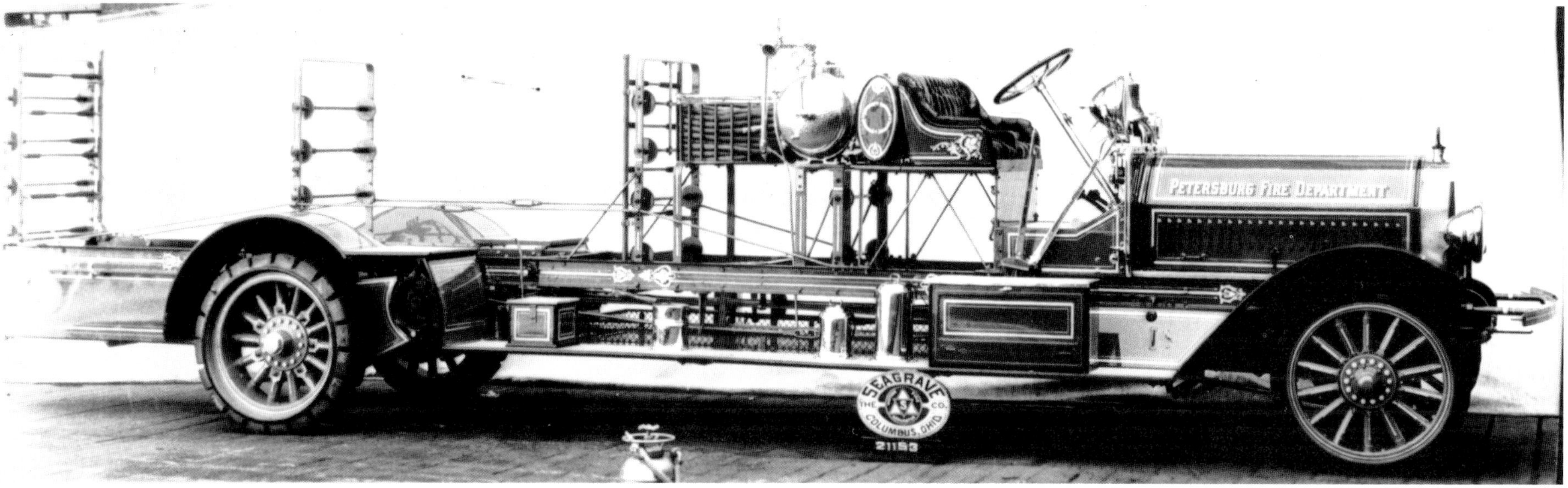
PETERSBURG FIRE DEPARTMENT
SEAGRAVE
THE CO.
COLUMBUS, OHIO

BRYN MAWR
FIRE CO.
B. M. F. C.
SEAGRAVE
COLUMBUS, OHIO

How Seagrave Trussed Ladders Are Constructed

There are 4 principal parts in Seagrave trussed ladders, namely: 1, Main (or heavier) Beam; 2, Truss (or lighter) Beam; 3, Struts (the short strips that hold the beams apart); 4, Rungs.

BEAMS: Both the main beam and the truss beam are of Douglas fir. The lumber for these is so carefully selected that, at best, only a few boards are obtained from a log 5 or 6 feet in diameter. The grain of the boards used in making Seagrave ladder beams must be straight and absolutely free from pitch pockets, knots, cross runs and everything else that would detrimentally affect the strength of the wood. These beams must not be from that portion of the log which is too close to the "sap" (or outer edge), nor too close to the heart. The utmost vigilance is used—at saw mill and at our factory—to detect any defects.

The seasoning of ladder beam lumber also receives every consideration at the hands of Seagrave experts, for it must be seasoned neither too fast nor too slowly, but to just that degree where the wood retains its great strength and lasting qualities.

No urgency, however great, would cause us to use lumber which had not been thoroughly seasoned.

STRUTS: Douglas fir is used for struts, and is as carefully selected and treated as that used for beams, because these short strips have their function to perform and they must do it correctly.

RUNGS: Seagrave ladder rungs are of carefully selected second-growth hickory, seasoned with as much care as the lumber used for the beams. Rungs receive at least five inspections, beginning with the boards as they come from the saw mill and ending with the turning and sanding. Rungs that show cross-grain, curls, or bad spots of any kind are discarded. It is not a matter of "this should serve its purpose," but, "this **must** serve its purpose."

ASSEMBLY: The two beams are securely fastened at both ends, and the struts are held in place by bolts—large enough to securely hold the struts, but small enough to not weaken them. Bolt heads are counter-sunk and nuts are rounded, to eliminate troublesome projections.

The curve of the two beams is that which experience has shown combines maximum strength with minimum weight.

Rungs are securely fastened in the **main** (or heavier) beam. That is an important feature of Seagrave ladder construction. By that method full advantage is taken of the tensile strength of both the main beam and the truss beam.

Under, and very close to, every fourth rung iron rods are securely bolted at each end to the main beams. This is done to give the ladder greater rigidity and more resistance to side stress.

You perhaps have seen ladders in which the rungs are set in blocks **between** two beams usually of the same size. In this construction there is a sacrifice of strength so far as the application

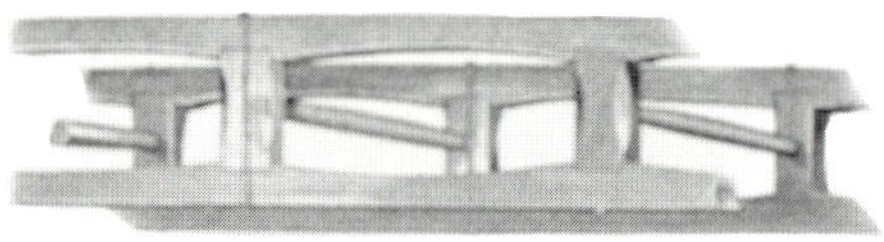

of weight to the beams is concerned. More than that, there is ever-present danger that the blocks holding the rungs may become loose through side stress, and split, because these blocks have a heavy bolt running through them. A double service is required of these blocks, for they serve as struts, (holding the beams apart) and they must support the weight borne by the rungs. It is apparent that the splitting of one of these blocks will result disastrously.

An advantage claimed for that type of ladder is that, because the rungs are in the center of the ladder, it may be climbed from either side. But this is a questionable advantage when strength and security are sacrificed thereby.

Seagrave trussed ladders may be climbed with either side up, and a Seagrave trussed ladder set with the truss (or smaller) beams **out** from a building will sustain as much, if not more, weight than any other type of ladder, however the greatest strength is available when our trussed ladders are set with the truss beams **in**, as shown in the illustration above. The placing of a ladder is such a simple operation—even in the haste of fire fighting—that firemen quickly become accustomed to placing ladders properly, both as to "front and back" and as to angle.

— BRANCHES —
Boston
New York
Philadelphia
Pittsburgh
Birmingham
Chicago

The Seagrave Corporation, Columbus, Ohio

— BRANCHES —
Kansas City
Dallas
Los Angeles
San Francisco
Portland
Seattle

Before the advent of fire doors, fire walls, sprinklers and other containment devices, scenes like this were common. It was easy to use up several hundred feet of ladders at an apartment fire. The wood ladder was a very durable life saver. But, it did require special care and some periodic maintenance. The following rules were published in many department handbooks.

- Never walk a ladder or pick it up by opposite corners.
  - both of these practices cause serious internal strains.
- Tighten all nuts every six months.
  - the most thoroughly seasoned wood shrinks with age.
- Never nail rungs.
  - if rungs loosen, pull the ladder apart and glue them, use only liquid furniture glue - never hot glue.
- New rungs should be placed with the grain vertical.
- Don't drop a ladder.
  - this loosens the rungs.
- Wipe off wet ladders. Keep well varnished.
  - moisture raises the grain of the wood.
- The distance from the wall to the ladder should be about one third the height of the ladder.
  - position a thirty foot ladder ten feet from the wall.
- Both stay poles on extension ladders should be the same distance from the base.
  - this is done to equalize strain.

To show the strength of the Seagrave truss style ladder this comparison was made. The Seagrave truss ladder, a standard 20 foot model, weighs only 65 pounds, but can support 1,642 pounds with only the slightest deflection. The individual weights of the men on the ladder were left to right: 142, 135, 151, 170, 180, 219, 185, 174, 150 and 136. The heaviest men were standing in the middle. The competing style ladder . . . well, the photo says it all.

The water tower for St. Paul, Minnesota, was ordered March 15, 1917, and delivery was to be before May 31. Well . . . they didn't make it. Delivery was on September 26, 1917. If the tower got off to a slow start it made up for it in longevity. It received a few updates over the years . . . like the steel wheels shown in this 1954 fire scene photo. Forty-four years after its delivery it was retired, but it still shows up at current musters to thrill the crowd.

The photos, at the right, are deliveries for Detroit, Michigan, and Chicago, Illinois. The 1924 Detroit delivery served until 1953. Chicago's 1925 Seagrave is still listed in active service, but is used only for parades.

Carefully studying these giants of the past one could see the evolution to the rear mounted aerial ladder of today.

TH COLUMBUS
SEAGRAVE
THE
CO.
COLUMBUS, OHIO

SEAGRAVE

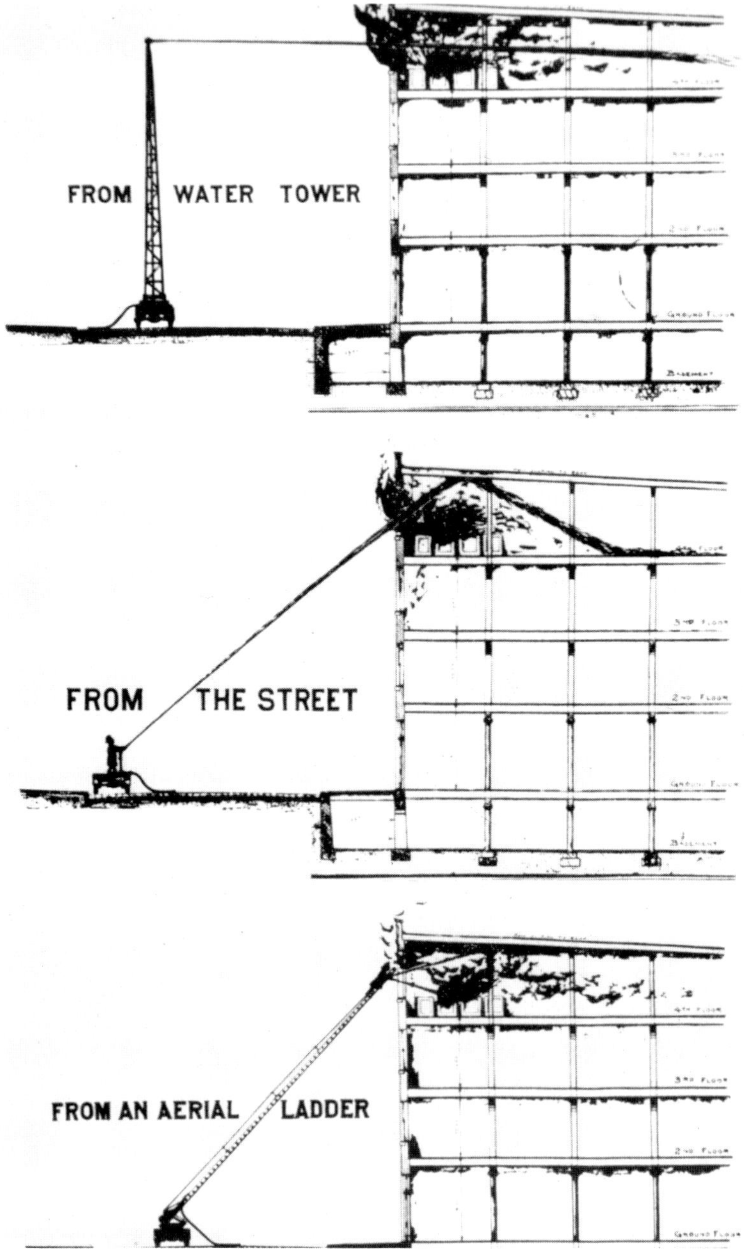

The water tower was an impressive work horse. Crowds always gathered when it was used, either at a practice drill or while in use. The Seagrave water tower always was used with a forward lean. That way the amount of lean could be used to counteract the wind or pressure generated by the stream. When these towers failed, it was usually when the mast was forced over center. Then the weight of the tower was added to the force of the stream.

Like the horse, just years before it, the water tower was doomed. Its technology, though sound, was no match for the sturdy aerial ladders now on the scene. There would be a few more deliveries to cities with many high rise buildings. But, practice drills down at the river, like the one shown at Chicago, Illinois, would soon be just a fond memory.

The Round Hood Era

1921 - 1935

First round hood pumper for the 1921 convention

The Round Hood at Seagrave was more than just a sheet metal style change. Refinements in engineering were concealed under this sleek skin. Starting in the boom years of the early 1920's, Seagrave continued to make advances in equipment design through the Depression and beyond. In fact, some of the largest design changes and development took place in the Depression years.

Motors became more powerful and efficient, with the advent of the V-12 engine. Aerial ladder mechanisms became simplified with the new gear box arrangement in 1927 and the all steel hydraulic ladder in 1935. Pumps, always centrifugal, changed in configuration and controls were modified. Driveline changes included the implementation of a shaft drive. All wheel brakes were adapted in 1928 and hydraulic brakes in 1931. Even the steering wheel moved from the right to the more conventional left-hand drive.

By the end of this era, the tires on the apparatus would completely change to all pneumatics, even a few white wall tires. Square-cornered, iron reinforced fenders would give way to smooth gull wing styles to complement the general streamlining process.

Seagrave added more buildings to the High Street location in 1928 and instituted their own Canadian factory branch, in 1929. Employment at the plant rose and fell with the economy. But there were now second generation Seagrave employees, and Seagrave was a more stable employer than most.

The smaller Suburbanite pumper was introduced in 1923. The Suburbanite filled a gap in the complete line of pumpers that Seagrave offered. The small equipment line would continue through the era on a commercial chassis, starting with a model A Ford utility platform aerial, in 1929.

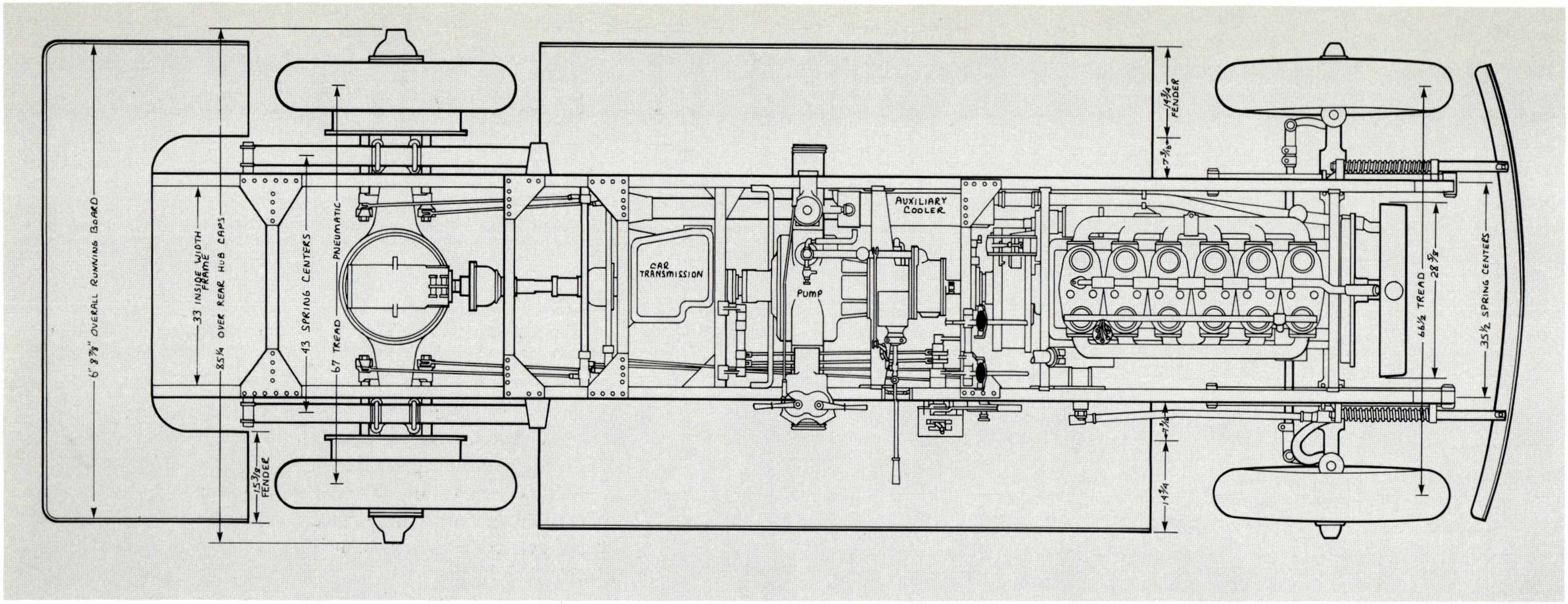

6' 8⅞" OVERALL RUNNING BOARD
33 INSIDE WIDTH FRAME
85¼ OVER REAR HUB CAPS
43 SPRING CENTERS
67 TREAD
PNEUMATIC
15⅜ FENDER
CAR TRANSMISSION
PUMP
AUXILIARY COOLER
14¾ FENDER
66½ TREAD
28⅝
35½ SPRING CENTERS

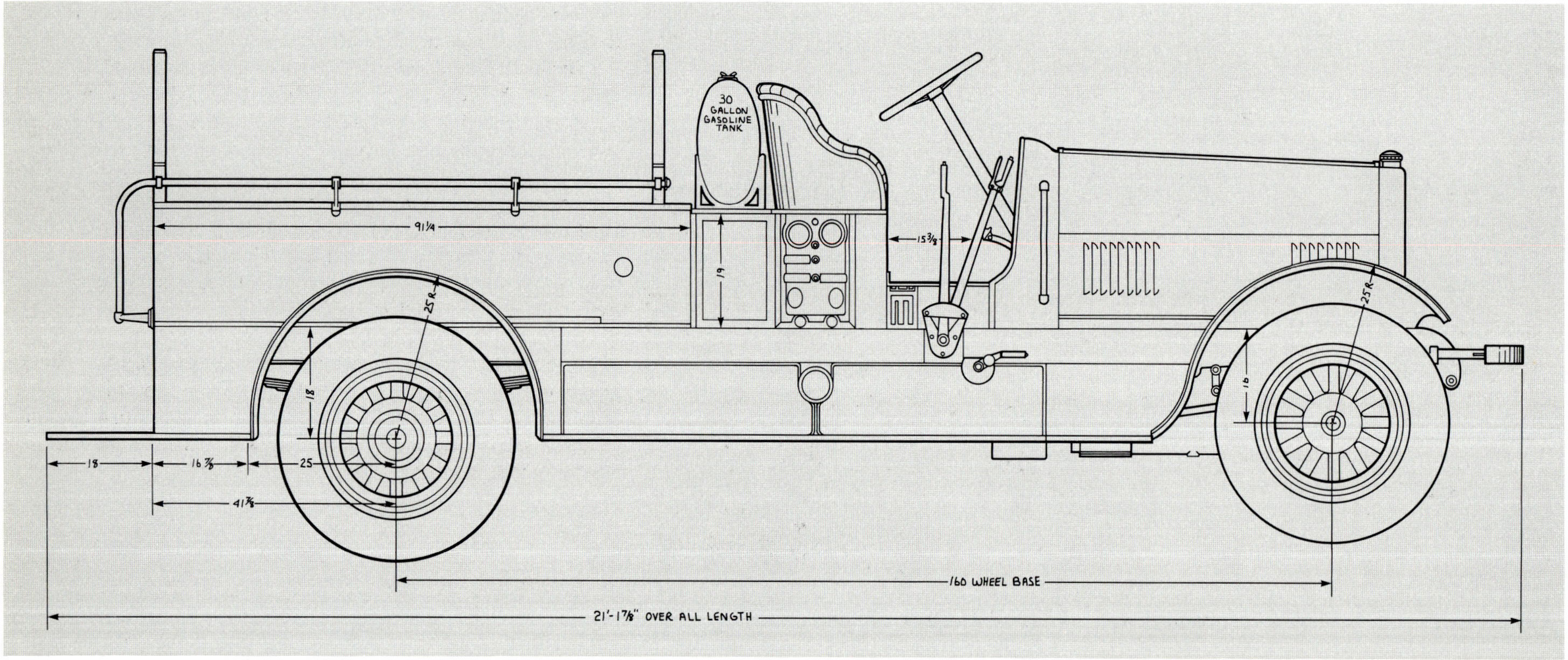

30 GALLON GASOLINE TANK
91¼
19
15⅜
25R
18
16
18
16⅞
25
41⅞
160 WHEEL BASE
21'-1⅞" OVER ALL LENGTH

These plant views were taken in 1926. The Seagrave style truss ladder was then 45 years old. In the motor room, Seagrave had been assembling engines for 19 years. Over in the pump shop, Seagrave built centrifugal pumps that were then 12 years old. In the aerial ladder assembly room, teams were building the new design ladder hoist mechanism.

The old buggy factory entered the Flapper Era without much change. Here in the main building, built in 1895, 14 inch square wood columns support the second floor where the machine shop was located. The upper floors were lighter in construction, as they were originally used for storage of buggy bodies while the paint dried.

Employment in the plant, in 1925, was 300. The plant was producing around 200 pieces of apparatus a year. For 1925, the painter's log shows 196, but not all deliveries were shown in the master painter's log.

Supplemental records show a local paint shop was doing some of the painting. The local shop received $190 to paint a standard pumper, but only $156 for a Suburbanite. A tractor trailer aerial price was $347 for labor and $109 for material.

The invention of Duco®, a quick-drying hard finish enamel, shortened painting cycles, but notes in the painter's log indicate "took less than 15 days" if a job was rushed.

The future looks bright for these Seagrave employees. Sales for the previous ten years showed an almost steady increase.

YEAR	SALES
1914	$ 702,000
1915	709,000
1916	884,000
1917	1,088,000
1918	1,262,000
1919	1,227,000
1920	1,416,000
1921	2,022,000
1922	1,518,000
1923	1,630,000
1924	2,062,000

Here is the true artist plying his trade. These decorators, as they were called, were given some license to be creative. Designs were allowed to vary, and work orders contained notes which only said "name of town curved over hood". The famous Indians that appeared in many forms on apparatus over the years all bear a striking resemblance to the painting on the wall.

October, 1921, saw the introduction of the new style round hood and cast aluminum firewall. The first rig through the paint shop, with these new features, was actually a rebuild of a wreck for Warren, Ohio. The second was a pumper for the 1921 Chief's Convention. The convention rig was sold to Pitman, New Jersey.

Start-up on the new design was slow, and intermixed with the gable hood models for years. The actual mass delivery of round hoods would be delayed until March, 1922.

The older gable style hood would remain popular, until 1926. Deliveries to Fayetteville, North Carolina, and Hull, Massachusetts, marked the end of a 14 year run of the gable hood models.

The old firewall, of a laminated wood base covered with oak veneer, gave way to cast aluminum. Luxury cars of the late teens and early 1920's made extensive use of cast aluminum, from steering wheel spiders to whole body sections. Seagrave was able to transform part of their brass foundry for making aluminum castings.

The round hood, and its accompanying open grille shell, would be adapted to all sizes of the Seagrave line up. From Metropolite to Suburbanite, the gentle curves of the hood denoted Seagrave's flair for style. The demand for a wide cab in the middle 1930's and general styling trends forced a change after almost 15 years of production.

Four sizes of pumpers would dominate the Seagrave line up through the Round Hood Era. First was the Standard 750/1,000 gpm pumper, like this 1925 delivery for Buffalo, New York. The F-6 engine powered these units, typical pump size was 750 to 1,000 gpm.

Lansing, Michigan, received this H engine powered, 1,300 gpm unit, termed a Metropolite, in July, 1925. Constant improvements in pump and engine design now allowed the 180 horsepower H engine to pump 1,300 gallons per minute.

In 1923, Seagrave introduced the Suburbanite series of apparatus. A full custom truck, but downsized, this 350 gpm model was driven by a Continental 6-B engine of 70 hp. The Atchinson Topeka and Santa Fe Railroad yards used this 1927 Suburbanite.

To round out the model line up, Seagrave introduced the Special series, in 1926. A Seagrave-Continental flathead six-cylinder engine of 105 hp powered this 600 gpm unit. Huntington, West Virginia received their classic Seagrave Special in December, 1929.

Canton, Ohio received this hose and chemical truck, in 1925. Straight chemical apparatus was on the way out by the mid 1920's. This was to be the era of the big pump. Although the photo of Canton, Ohio's, rig is over fifty years old, its life saving message is still apropos.

The basic engine of this period was the F-6 motor of 1,013 cu. in. displacement. The two less common motors were the H-6 of 1,717 cu. in. and the F-4 at 675 cu. in. displacement. The F-6 was in 85 percent of the large pumper deliveries. The H-6 had 10 percent of the market and the F-4 had the balance. Twenty six louvers on the side of the H-6 hood indicated the large engine vs sixteen louvers for the F-6 and ten louvers for the F-4.

A custom builder truly builds custom trucks. Compare the intake casting on this 1927 delivery for Richmond, Virginia, with the Lansing, Michigan, delivery on the previous page. The variation of rear inlets, hose box size, and overall dimensions dictated that the engineering staff was involved with every delivery. Lansing, Michigan, ordered Sewell cushion wheels with Kelly Springfield air core caterpillar tires. Richmond, Virginia, ordered disc wheels and Overman cushion tires.

Both of these H motor monsters would pump 1,300 gpm with a 10 foot draft and a motor speed of only 920 rpm. The H engine is shown on page 199.

Superior, Wisconsin, and Chicago, Illinois, appeared to have twins. Superior received a standard unit, while Chicago dictated fourteen mechanical changes and another dozen appearance changes. Changes included: a hand lever between the driver's and officer's seat for operation of the flywheel mounted siren, red and green headlights, and the bench seat over the hose box.

C.F.D.
SEAGRAVE
THE CORP.
COLUMBUS, OHIO
53170

SEAGRAVE
THE CORP.
COLUMBUS, OHIO
68805

Quincy, Massachusetts, received this 750 gpm pumper in the closing days of 1925. This pumper had a 750 gpm pump as opposed to the Charleston, South Carolina, pumper which has a 1,000 gpm pump. The difference is noted by the pump discharge handle and its location toward the edge of the front seat. The bigger 1,000 gallon pump sat farther back in the chassis. Compare the pump location to the 1,300 gpm pumper on page 156.

Pump impeller diameter remained standard, but the pump to engine rpm ratio could be changed to produce a larger volume pump. The rule of thumb is 250 gpm discharge through each outlet. Outlets are only one factor, others include engine size, suction hose size and lift of water. Many times a delivery man would point the end of a suction hose upstream, in the coldest stream he could find, trying for the best performance.

Shown in this, May 6, 1926, photo, at a pump test, are a 400 gpm Suburbanite and a 1,000 gpm Standard pumper.

The Charleston, South Carolina, 1930 delivery features a preconnected "squirrel tail" suction hose and a CO^2 powered chemical delivery system. The chemical system was powered by a gas cylinder under the tank (similar to the rig on page 155). Recharging the gas cylinder was cumbersome and the system was soon dropped.

The wrap around, preconnected suction hose, on the other hand, saw favor in many communities for another 30 years. Few big cities with good hydrant systems appreciated the convenience, but many rural communities favored preconnected suction hose.

Driving and pumping with one of these 1920's vintage Seagraves is delightful. When you "pop" the clutch on a 1,000 plus cubic inch, long stroke, gas engine, you can feel the power. The front end will start to lift and the roar of the 3 1/2 inch exhaust will signal excitement.

Many of these vintage pumpers will still pump at capacity today. As water passes through the pump, it actually polishes the pump. The polishing reduces turbulence and increases flow. Many Seagrave bronze pumps actually produced five percent better flow a year after delivery.

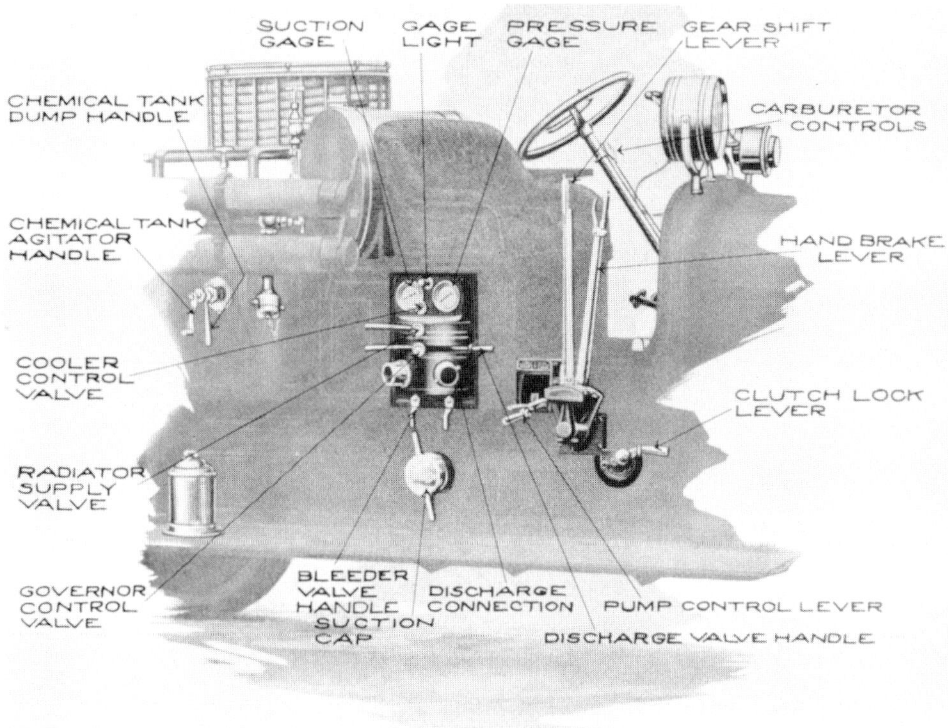

SACRAMENTO
2
SEAGRAVE
THE
CO.
COLUMBUS, OHIO
37350

C. F. D.
SEAGRAVE
THE
CORP.
COLUMBUS, OHIO
57911

Note the changes in engineering and use from the Sacramento, California, chemical rig, in 1924, to the Cleveland, Ohio, 1,000 gpm pumper, in 1929. Sacramento's order, in 1924, was for three 750 gpm pumpers and this last of the straight chemical cars. The chemical tanks were 50 gallons each. A 2 1/2 inch fitting allowed these tanks to be refilled at a hydrant, or hydrant pressure could be transmitted through the booster hose.

The most obvious mechanical changes were the change from chain to shaft drive, bumper style and the change to Budd ten hole disc wheels. The disc wheel allowed room for brake drums. Detroit, Michigan was the first to use the four wheel brake apparatus. Seagrave used a "D" in the model designation for the ensuing years to indicate four wheel brakes.

The delivery of an apparatus was cause for a crowd in many towns. Here, the towns people of Menominee Falls, Wisconsin, greet their new pumper, in 1924. Most deliveries were by rail. Even towns as close as 100 miles to Columbus, Ohio received rail deliveries.

A letter or telegram would be sent saying, "We have instructed C & O (or some other rail line) to deliver car #XXXX to the rail siding in your town with the end opening car facing East." End opening cars came in 36, 40 and 48 foot lengths. Aerial ladder trucks were shipped in two separate box cars.

This photo of the Grosse Pointe Park, Michigan's, 1928 Seagrave graced many Seagrave advertisements. The Grosse Pointe's were a collection of affluent Detroit, Michigan, suburbs to the North and East of the city. They could afford the best, they chose Seagrave. For many years, "The Grosse Pointes" were the hometowns for the auto giants. The Fords, Dodges, Fishers (Fisher Body), all had estates protected by Seagrave apparatus.

Montreal, Quebec, was a Seagrave user since delivery of its first water tower in 1905. Serial #56386 was a 1929 delivery. Bilingual Montreal had "M.F.D." on one side of the hood and "S.I.M." (Service Incendies Montreal) on the other. Montreal ordered chain drive apparatus as long as they could because the lack of a differential housing gave the apparatus more ground clearance in the snow. The high ladder mountings provided extra hand rails. Fifteen of these engines were ordered from 1921 to 1931. This restored truck is the pride of Eric Sprenger, long time Seagrave historian.

A U. S. twin to the Montreal, Quebec, delivery was serial #53165, a 1928 delivery to Virginia, Minnesota. The pumper is still in service today at a resort in Northern Minnesota, from April 1st to November 1st each year. Receiving only 3,000 road miles since new, it remains the most original early Seagrave.

The town of Virginia, Minnesota is heated by a central steam heating system. The lack of a fireplace or furnace in each building in town accounted for the low usage of the fire engine.

Fireplace fires kept many an early department busy. Flames or flying embers could blow back into living spaces or find their way through cracks in the chimney to roof members.

Furnaces had problems of their own, with overheating and loose fitting joints, topping the list. And then there were hazards with fuel storage and ash removal. Soot ash and smoke were common with individual furnaces. It would not be uncommon for a child to walk to school on a blanket of white new fallen snow, and return home to see the blanket turn grey.

Twice floods kept this engine on extended service. Pumping for 100 hours straight kept flood waters from reaching the main powerhouse boilers in the early years and a similar session kept the telephone office dry, in 1952.

Less than 10 percent of Seagrave deliveries through the 1920's were powered by Seagrave's F-4 engine. The F-4 was rated at 53 S.A.E. horsepower, as opposed to 79.3 for the six-cylinder engine. The Brockton rig was powered by an F-4 engine as noted by the short hood. This truck also featured a 40 gallon chemical tank and the famous Springfield hose body. The wide body provided an area for added bench seats.

The delivery for Grosse Pointe Shores, Michigan, featured a Morrow metal wind breaker. The sheet metal shell contained celluloid inserts for the driver and officer. In later years, as the celluloid turned yellow, it was replaced with plexiglass. Morrow wind breakers were a $100 option. Some departments removed them in summer, but they were welcome back in the fall as the temperatures dropped.

Springfield, Massachusetts, ordered these two rigs only two years apart: the chemical in 1926, and the squad truck in 1928. Both bore the distinctive black paint with gold trim. Serial #43484 had right-hand drive and the F-6 engine, as indicated by the sixteen louver hood. The squad rig, with the twelve louver hood, is a left-hand drive with the special six-cylinder engine built by Continental. This squad car was the 25th delivery of motor fire apparatus to the city of Springfield.

S.F.D.
SEAGRAVE
THE CORP.
COLUMBUS, OHIO
53575

SEAGRAVE
THE CO.
COLUMBUS, OHIO
43484

All Seagrave Specials were rated at 600 gpm and had two discharges on the driver's side and one on the curb side. Only a rear intake was standard, except on a combination ladder truck. These trucks featured a hose body with a capacity for 1,200 feet of 2 1/2 inch hose and four-wheel mechanical brakes. The option list on the Specials remained smaller, so they would not inhibit sales of the bread-and-butter Standard series.

A dramatic difference in styling can be seen between these two squad cars. On the right is a May, 1930, delivery for Los Angeles, California. Above is a December, 1934, delivery for Bakersfield, California.

Records show employment at Columbus, dropped almost in half from 1930 to 1934. Some of the 175 employees listed in 1934 only worked two or three days a week. But the design staff was busy. This new wide hood and cowl section were needed to install the Seagrave built V-12 engine that debuted at the low point of the depression, 1932. Rolled fenders, fork mounted headlights, and right-hand drive were things of the past.

Under the hood, the F-6 engine received an update. The flywheel was now enclosed so that the F-6 and V-12 could share a common transmission mounting. The V-12 was actually smaller in displacement than the old six-cylinder engine (906 versus 1,013 cu. in.), but both weighed about the same.

By 1931, New York City had received 108 deliveries from Seagrave, but not one pumping engine. The city, until now, had been content with motor pumpers from Ahrens-Fox and American LaFrance.

The city purchased nine, special 600 gpm pumpers on August 13, 1931, and all were delivered within ninety days. Power for these pumpers was provided by the new Seagrave, designed and built, G engine. The new engine had all the cylinders cast in one block, as shown on page 198.

The new pumping engines were very easy to drive and operate especially for the engineer who previously had an Ahrens-Fox. Hard rubber tires were almost a thing of the past, but they were still listed on New York City's specification sheet. Hard rubber tires were replaced by pneumatic tires, in the 1940's, as pumpers came in for repaint. The fleet survived well into the 1950's as spares.

In 1932, New York ordered eight 1,000 gpm pumpers. This time, power was supplied by the venerable old F engine. This order was to include five combination pump and chemical engines, but the depression took its toll and only the straight pumpers were delivered.

The separate cast cylinders of the old style F series engine required more room than the single cast G-6. That's the reason the hood is longer on these deliveries.

In 1933, New York City ordered five hose wagons. The wagons were powered by the G engine and featured vacuum assisted brakes. One assigned to Engine 65 was lost in a fire on January 15, 1942. The four survivors each served for over twenty years.

The firemen of New York City liked the leather straps on the previous year deliveries. The subway style strap was reordered for the 1933 deliveries. The strap changed to solid metal in later years and became a New York City trademark until the mid 1960's. But, times change, earlier this year one of the straps was found at the N.Y.C. shops and few employees could identify it.

The photo, to the left, is one of the 1931 600 gpm models (as described on page 168) after conversion to pneumatic tires. Engine 163, serial #67638 was the only one of this model to carry a 24 foot extension ladder in addition to the standard F.D.N.Y. scaling ladders.

Ten of these handsome pumpers were ordered in 1934. Eight on a standard order and two to replace two supposedly rejected American LaFrance pumpers. One of the 1934 deliveries, serial #77126, Engine 70, was the first pumper to get a two-way radio (in 1940).

In 1932, Seagrave tried to spark a little extra interest by enhancing product designations.

The **Standard** pumper stayed the same: 1,000 gpm, four stage pump, powered by the F engine of 5 3/4 x 6 1/2 inch size, shown on this page.

The **Sentinel** pumper was a 1,000 gpm, four stage pump, powered by the G engine of 5 1/8 x 5 3/4 inch size, as in the Cambridge, Massachusetts, delivery.

The **Sentry** pumper was a 750 gpm, single stage pump, powered by the G engine of 5 1/8 x 5 3/4 inch size, as in the Pittsfield, Massachusetts, delivery.

The Standard pumpers, shown, are for Akron, Ohio, and Tiffin, Ohio. The Tiffin unit is shown on the day of retirement from active service, in 1972. Delivery time for this apparatus was only four weeks from the date of order, which shows how slow the plant was during the Depression. There were 46 deliveries in 1934, and six of these were repairs.

Today, the truck from Tiffin is still a marvel. The sight and sound of this truck at parades and musters awes the crowds. The throaty sound of the 1,013 cu. in. engine under load can be heard for blocks - no siren needed.

P.F.D.
3
SEAGRAVE
THE CORP
COLUMBUS, OHIO
77990

CAMBRIDGE
5
SEAGRAVE
THE CORP
COLUMBUS, OHIO
78810

As the Depression deepened, sales dropped. Only 48 orders in 1933, and 13 of those were to New York. The sales volume may have dwindled, but the finish of the equipment remained at a high level. The rigs shown, are a 1932 delivery to Akron, Ohio, and a 1935 delivery to Cambridge, Massachusetts.

The G engine was now at the same basic horsepower rating as the venerable old F engine. But the G engine was more reliable, lighter, and less costly to build. Less weight up front meant more speed and maneuverability for these rigs.

Restoring any vehicle is a task, but fire engines are an elusive quest. For example, both of the pumpers on this page have the same dashboard, but a different set of gauges.

The 1924 delivery to Houston, Texas, Engine 24, was sideswiped by a street car, in 1934. One fireman was killed in the accident. Seagrave rebuilt the apparatus, in 1935. The result, a rig with a 1935 serial number, but left hand drive, wood wheels and chain drive. What's a restorer to do?

24

FIRE DEPARTMENT
24
SEAGRAVE
THE CORP.
COLUMBUS, OHIO
51192

Somerville, Massachusetts placed an order, in 1932, for two pumpers, two hose and chemical cars, a service ladder truck and two very unusual pieces of apparatus. They were a custom wrecker and a search light truck.

Close study of the simple Seagrave design would reveal the basis for modern recovery vehicles. Notice the absence of overhead framing and support cables, just a boom supported from the bottom. Full size wreckers were operated by many departments.

Somerville, also received this search light car, in 1932. A four-cylinder gas engine powered a Koehler generator located in the forward storage compartment.

Los Angeles, California's hose wagon was unique in its own right, with that Greenberg monitor. Most truck mounted monitors or water cannons were not equipped with hand wheels for rotation and elevation.

FIRE DEPT.
SEAGRAVE
THE CORP.
COLUMBUS, OHIO
70051

ENGINE COMPANY
SEAGRAVE
THE CORP.
COLUMBUS, OHIO
45727

SEAGRAVE SUBURBANITE

Fire Department Pumpers, Chemical Cars, Chemical and Hose Cars, City Service Ladder Trucks

THE SEAGRAVE CORPORATION — COLUMBUS, OHIO, U. S. A.

The main attraction at the International Association of Fire Engineers, in 1923, was the new, smaller Seagrave, the Suburbanite. This low priced, lightweight fire engine was a full custom truck. It could be ordered with a 350 gpm pump, chemical equipment, squad or a combination of the three. (For information on Suburbanite ladder trucks, see page 194.)

Power was provided by a Continental 6-B engine until 1931, when a switch was made to a Hercules YXC engine. Overall height was less than six feet, and so was the width. The approximate weight was 6,700 pounds. Small towns, industrial plants and big cities all saw the value in these efficient units.

Detroit, Michigan took delivery of two pumpers and two service ladder trucks, in 1925, and two more ladder trucks, in 1926. They were used where the roads were poor in outlying areas. They were priced right, at $6,684, so they could compete with top-of-the-line commercial pumpers.

The four deliveries on this page were all made in the summer of 1930 and are all a little different.

Serial #		
61825	July	Mount Gilead, North Carolina
61986	August	Walhalla, South Carolina
62265	June	Westfield, Massachusetts
61535	May	Greencastle, Pennsylvania

FIRE DEPT.
1
SEAGRAVE
THE CORP
COLUMBUS, OHIO

RESCUE HOSE CO.
No.1
SEAGRAVE
THE CORP
COLUMBUS, OHIO

F.D.N.Y.
SEAGRAVE
THE CORP.
COLUMBUS, OHIO
57203

On December 19, 1929, at a city auction, the last remnants of New York City's horse-drawn era, a chief's wagon was sold. It brought two dollars. That same year New York City ordered 50 Suburbanite hose wagons. The following year 23 more Suburbanites were ordered. These hose wagons had to follow around pumpers of other makes because New York City didn't order Seagrave pumpers until 1931. Even the officer's side of this F.D.N.Y. delivery has a rugged appearance.

The 1931 delivery to Stoneham, Massachusetts, had the upgraded driveline with a 94 hp Hercules engine and new style pump.

Shown next to Frederic Seagrave's final resting place is an ex-Norfolk, Connecticut, 1925 Suburbanite. After 15 years in service, it was sold to the General Electric Plant in Philadelphia, Pennsylvania. It is now owned by fire apparatus historian Walter McCall and is stored just blocks away from this cemetery in Windsor, Ontario, Canada.

BAMBERG
FIRE DEPT.
SEAGRAVE
THE
CO.
COLUMBUS, OHIO
38890

WESTOVER
FIRE DEPT.
1
WESTOVER F.D.
SEAGRAVE
THE
CORP.
COLUMBUS, OHIO
46695

These photos illustrate the changes that took place over the life span of the Seagrave Suburbanite. Shown here, are serial #38890, a 1924 delivery for Bamberg, South Carolina; serial #46695, a 1926 delivery for Westover, West Virginia; serial #63140, a stock sales car for the Los Angeles branch in 1930; and serial #80000, a 1935 delivery for Las Vegas, Nevada.

The early Suburbanites came with painted headlights by Brown Lamp Company and with no bumper. That was typical of the spartan commercial trucks of the day. The shape and style of the headlights reflected the style trends.

All early Suburbanite pumpers had front intakes, the later deliveries side intakes for the pump (compare the photos here with page 178). By the mid 1930's, the sheet metal had changed, but the mechanical components and the name remained . . . Suburbanite.

24
SALVAGE AND RESCUE SQUAD
ORANGE
VOLUNTEER FIRE CO.
O.F.D.
SEAGRAVE
THE CORP
COLUMBUS, OHIO

McSHERRYSTOWN
F.D
2
Mc SHERRYSTOWN
STEAM FIRE ENGINE Co.
No.1
SEAGRAVE
THE CORP
COLUMBUS, OHIO
72345

The classic, depression era, Suburbanites show how adaptable the basic chassis was. Shown here, are serial #66445, a 1931 delivery for Orange, Virginia, and serial #72345, a 1933 delivery for McSherrystown, Pennsylvania. The Suburbanite was certainly better trimmed than the average commercial fire truck of the day.

The Suburbanite was a forerunner of the way independent truck makers build today. This pumper was assembled with the best components available in a custom cab and frame. The Seagrave six inch frame provided the nesting area for: Timken axles, Bendix three shoe mechanical brakes, Ross steering gear, Brown-Lipe #31 transmission, and a Delco ignition system.

The ads for the Suburbanite boasted: "The centrifugal pump is guaranteed to deliver rated capacity, barring abuse, for 10 years." This single stage, one impeller pump would even out last the Suburbanite model. In a slightly modified form, it stayed in production over 25 years. Shown above, are the early and late model Suburbanite pumps.

The early Suburbanites with the 6-B engine were joined with the four stage pump, top left. The later more powerful engine would produce more pressure and capacity with the improved single stage pump. With the 1923 through 1928 deliveries came the Continental 6-B engine. This 3 3/4 x 5 inch six-cylinder engine was rated at 70 horsepower. Pump tests show that the engine and pump combination was at its maximum at 400 gpm. Test after test showed 400 gpm, 402 gpm, 404 gpm. Most Seagrave pumpers of the day would put out 15 percent above capacity.

With a switch to the Hercules engine of 90 horsepower and a modified side suction pump, ratings ever increased to 500 gpm, and pump tests showed 550 gpm or better. Of course, the new engine sported a carburetor, with a 1 3/4 inch throat vs 1 1/2 inch for the old 6-B.

Most engines of the day were simply poor breathers. When the up draft style carburetor was replaced, with a more efficient down draft style, a 10 percent gain in horsepower in the same engine was almost assured.

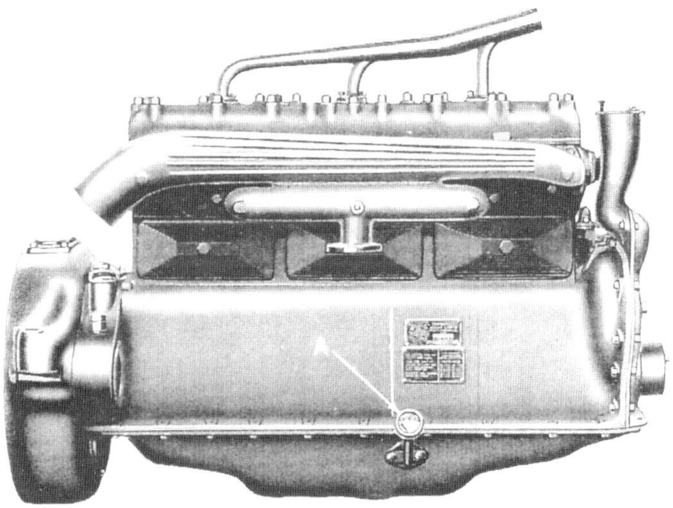

To offset slow sales during the depression, Seagrave mounted the single stage Suburbanite pump in a commercial chassis. Reo and Ford were the most common deliveries. The Reo Seagrave was referred to as a Fire Master and the Ford, a Volunteer. A 1935 Chevrolet, with the 3 5/16 x 4 inch six-cylinder engine, "the classic stove bolt six" could barely tweak out 500 gpm.

The Chevrolet was not a favorite pumping engine, until 1937, when its base horsepower was increased from 68 hp to 78 hp. The 1937 model Chevrolet also had four main bearings vs the previous three main bearing design.

Commercial chassis pumpers were favored by many small town departments because of their lower initial cost and ease of obtaining replacement parts.

From 1929 until 1935, Seagrave operated a plant at St. Catherines, Ontario. This plant acted as import agent and final assembler for the Columbus built apparatus. The peak employment at the plant was eight. This apparatus for Lacombe, Alberta, was fabricated at the plant, in 1932.

PORT ANGELES
FIRE DEPT.
PORT ANGELES, WASH.
SEAGRAVE
THE CORP
COLUMBUS, OHIO
82525

SEA CLIFF
L.I.
SEAGRAVE
THE CORP
COLUMBUS, OHIO
64090

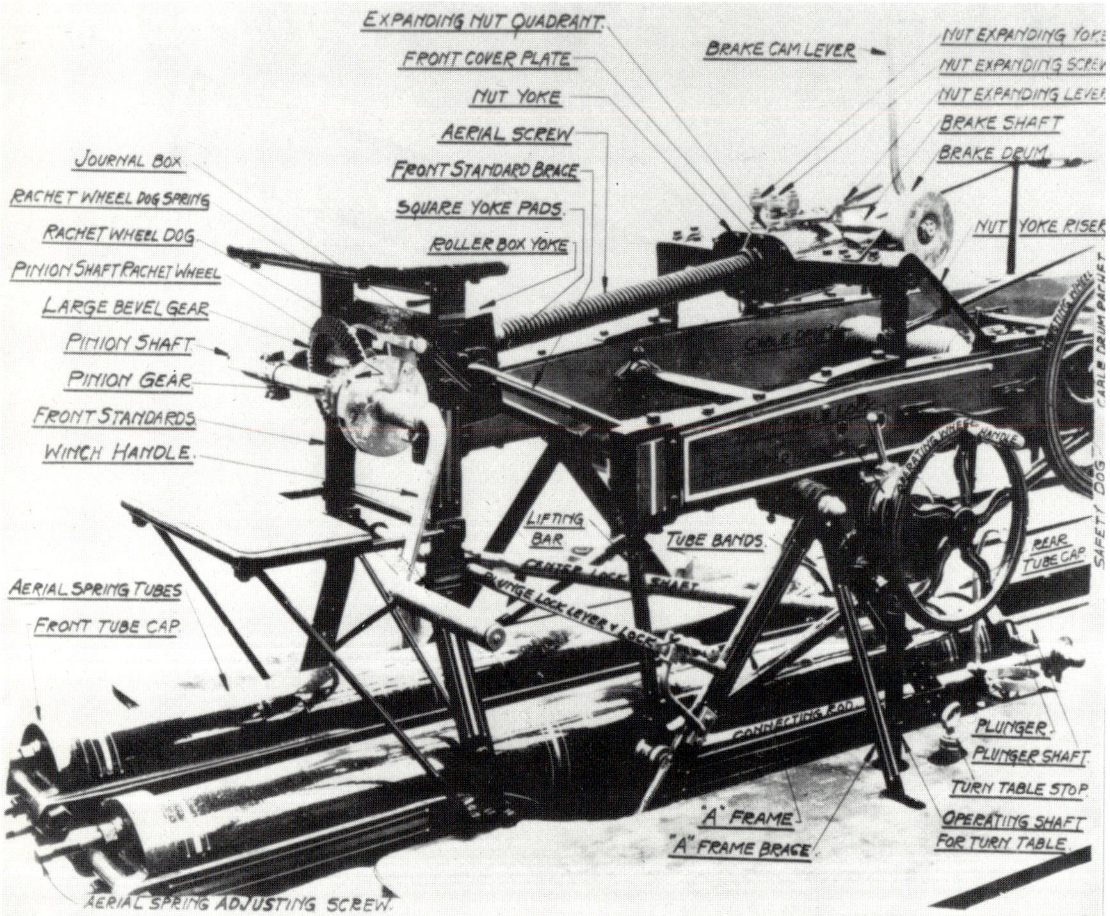

The aerial ladder was revamped, in 1927. The solid beam construction and spring hoist mechanism remained basically the same. The big changes were: moving the main screw between the spring cylinders, enclosing the entire mechanism and mounting the ladder under the pivot point.

The basic ladder configuration, now over 20 years old, was a proven commodity: no compressed air to leak like Ahrens-Fox, and no heavy mechanical arrangements like American LaFrance or Mack - just simple spring raise and hand crank down.

Placing the screw in the lifting frame not only made access easier for climbing the ladder, it took the strain of lowering the ladder off of the ladder beams.

Moving the ladder under the pivot point allowed lowering the center of gravity and the height to the tiller seat by ten inches. The frame of the trailer was now goosenecked ten inches to provide room for the usual complement of ladders. Height to the top of the tiller wheel was now nine feet.

Another change was to a solid casting instead of hand crafted iron for the aerial pivot platform. The wood beam of the main ladder was reinforced 12 feet from the lower end by a 1/4 inch thick pressed steel channel, 7 inches deep and 2 3/8 inches wide. These would be the final changes in wood ladder mechanism; the all steel aerial would arrive in ten years. The apparatus shown, are a 1927 delivery to New York City, and a 1928 delivery to Syracuse, New York.

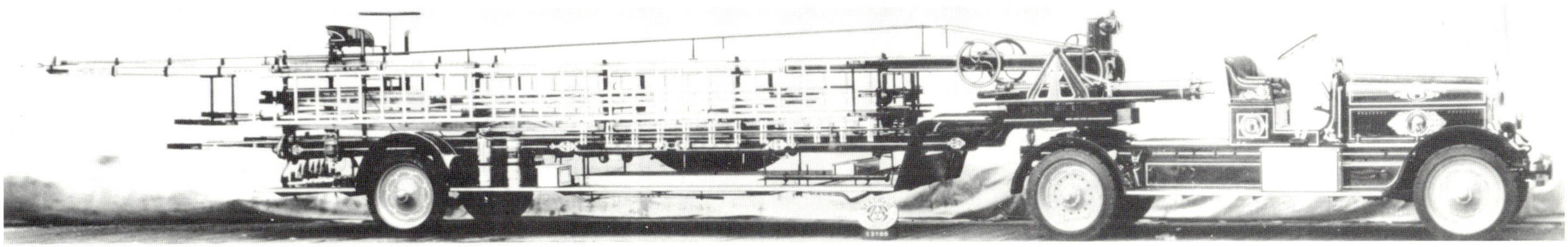

Shown above, is a February, 1929, delivery for Rye, New York. This 75 foot new style aerial would weigh 18,700 pound and have an overall length of 57 ft. 8 in. (an 85 ft. aerial would be 19,800 pounds and 62 ft. 8 in. long). Other key dimensions would be: a 27 ft. 6 in. turning radius, 21 in. to the running boards, a width of 7 ft. 2 in. and an overall wheel base of 46 ft. 3 in. (85 ft. aerials had an extra four feet of wheel base).

The trailer axle was now interchangeable with that of the steering axle of the tractor. The steering drag link on the trailer would work on a fore-and-aft motion, not crosswise as in previous models.

The photo, at left, shows the new rocker mechanism, or fifth wheel. Compare this basic style with the tractor on page 74. Note that the fixed or solid plate is now on the tractor. This feature eliminates the rock and sway that develops when the tractor and aerial are jackknifed at a fire scene. When at 90º angles to each other, the fifth wheel holds the aerial from tipping. The rotation of the aerial platform is enhanced by multiple rows of ball bearings placed in grooves in the fifth wheel.

75 foot aerial for Buffalo, New York - June, 1925

75 foot aerial for Portland, Maine - April, 1929

Service ladder for Los Angeles, California - November, 1929

The capitol, Washington D. C., favored tillered ladder trucks, as represented by these 1924 and 1925 deliveries.

The photo at left, illustrates the coordination that driver and tiller operator must maintain. The tiller driver must think of the back portion of the apparatus as his own car. He must envision that the grille for his vehicle is about ten feet ahead of his seat. When making a sharp right turn, for example, the tiller driver must not allow his imaginary grille to be pulled around the corner. He must steer a little left first (this will overcome the vehicle's tendency to be pulled into the turn), then he must turn a hard right, to swing the trailer, when his imaginary grille is at the corner.

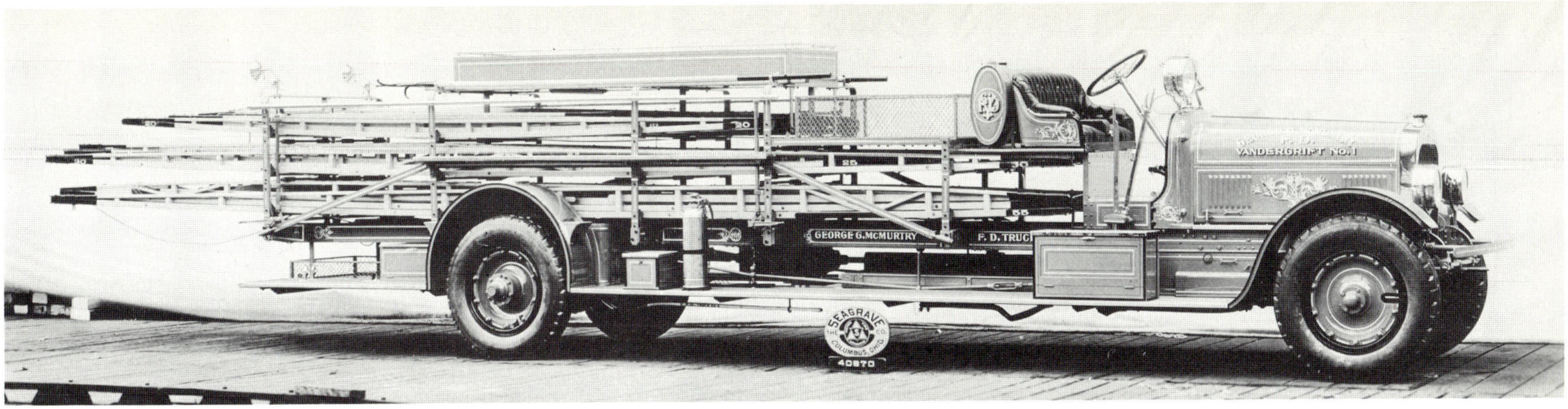

Service ladder truck sales peaked in the late 1920's. The reliability of the aerial ladder and the versatility of a ladder and pumper combination, spelled the demise of the service ladder rig.

Pictured on this page:

Serial #40870	Vandergrift, Pennsylvania
Serial #61510	Burlingame, California
Serial #63485	Lake Forest, Illinois

Standard equipment for a Suburbanite ladder truck was 219 feet of ladders. Extension ladders of 45, 35 and 16 feet; wall ladders of 22, 20 and 18 feet; and roof ladders of 16 and 12 feet were standard. Suburbanite ladder trucks left the plant at a little over 7,200 pounds. By the time the community loaded them up with equipment, they often exceeded five tons.

Communities, both large and small, ordered them. At the left, is a 1924 delivery to Detroit, Michigan. Above, is a 1930 Suburbanite service ladder for Livermore Falls, Maine.

The Suburbanite was the smallest of the Seagrave ladder trucks and at a price of $5,000, they were a bargain. Suburbanite ladder trucks with their six inch channel beam frame and four wheel brakes, were well suited to the poor roads of most suburban communities.

The largest of the Seagrave service ladder trucks was the standard Quad, shown at the right. The Quad's wheel base was 272 inches, a full 56 inches longer than the Suburbanite. It's ironic the little Suburbanite had four wheel brakes, but the large Seagrave did not have them, until 1928.

Combining a 750 gallon pump and a service ladder truck gave budget minded communities a full service fire department on one chassis. This combination sold well.

Ashland, Pennsylvania, received their truck, in 1924, and the River Edge, New Jersey, delivery was in 1930. This six year period brought subtle changes to the running gear. Most notable was the ten hole Budd disk wheels concealing the drum brakes. A booster tank and reel, along with soft suction hose, indicated that River Edge, New Jersey, had a very progressive fire department.

Seagrave was the most elaborately decorated apparatus of the era. Note how the hood center reflects the name of the community - even the frame and splash panels enhance the massive look of these trucks.

The dashboards contained: an ignition selection switch - far left; a speedometer - far right; an oil sight gage - bottom center; and an instrument cluster. An ignition switch, a light switch, an amp gauge and dash light shared the cluster.

Service ladder trucks tend to serve a long time: forty or more years is not uncommon. Most are rebuilt and modified to enhance their life. Shown above are: serial #38582, a 1924 delivery to Oakland, California, and serial #36565, a 1924 delivery to Denver, Colorado.

The photos, left, depict a rebuild of a service truck for Ansonia, Connecticut. The August, 1916, delivery was rebuilt in 1926 as a match for the new 750 gpm pumper ordered that year. These photos show the popularity of service ladder trucks coast to coast.

Seagrave produced a light aerial platform, for public utilities, from 1929 through the depression. The unit featured a wood and steel ladder with an 18 foot 4 inch reach, mounted on a swivel base.

The ladder used steps, rather than rungs, and an internal linkage allowed the 8 inch wide steps to always be horizontal, regardless of the reach of the ladder. The swivel feature allowed the truck to be parked next to streetcar tracks and remain stationary; only the ladder needed to be moved or rotated as the streetcar passed by.

The platform was 28 x 52 inches and made of oak. This wide stance made tree trimming and street light cleaning easier. The ladder beams were made of select white ash 1 1/2 x 4 5/8 inches in cross section. The wood sections were reinforced with steel. Supports were riveted to the pressed steel turntable which was bolted to the truck frame.

Seagrave would factory mount the platform, but preferred to sell the platform separately. Local communities could then mount the unit on existing trucks or modified passenger car chassis. Many large prestige vehicles, like Packards and Lincolns, found a second lift as municipal utility vehicles. The large heavy driveline would become the base for a wrecker or maybe a Seagrave aerial platform.

Continental - Special

Seagrave - Sentry or "G"

The basic power plants for the Round Hood Era were the Seagrave built F and H engines. They were designed my Homer "Cap" Havens, formerly with Frayer-Miller. This Ohio State University engineer was impressed with the Gorham engine, and he adapted and improved it for more fire apparatus use.

These series of engines were built in both four and six-cylinder versions, but the six-cylinder dominated 90 percent of the mix. The H engine powered the 1,000 plus gpm Metropolite, and some larger ladder trucks.

The F engine, first introduced in 1915, went through twenty or more revisions. Early build specifications called for the 19th series parts list, for example. The engine was sold for other uses also; the city of Honolulu ordered F engines for its fire boat.

The first F engine delivery went to San Diego, California, with serial #11678, in 1915. It featured separate cylinder castings of cast iron on a bronze crankcase. The last F engine had chrome nickel alloy castings on an aluminum crankcase. It was delivered to Jackson, Mississippi, in 1937.

Optional engines of the Round Heed Era were the 6-B, shown on page 185, and the Special and Sentry show here.

The engine, shown at the far left, was the Seagrave Special engine. It was a six-cylinder, 4 1/2 inch bore and 5 3/4 inch stroke design with 105 hp, built by Continental. It was used in the Seagrave Special, or mid range pumpers through the 1920's and 1930's.

The Seagrave built G engine was a 5 1/8 inch bore and 5 3/4 inch stroke design with 168 hp. This durable power plant was an option in the large pumpers and ladder trucks from 1931 to the 1940's. The last G engine was in a ladder truck for Milford, Massachusetts, in 1941.

The Seagrave designed F and H engines were engineering marvels. The F engine was of 1,013 cu. in. displacement, and conservatively rated at 79 hp. The H engine was of 1,717 cu. in. displacement and rated at 109 hp. Each cylinder of the H engine, at 286 cu. in. was bigger than some whole automobile engines today. The total volume of the H engines was 28 liters. The comparative size of these engines becomes evident under the hood or on the test stand.

Innovation

Innovation was always a key word at Seagrave during the Round Hood Era. Seagrave had perfected the centrifugal pump, updated the wood ladder, and created a whole new line of pumping engines.

Some changes were small, like the hose tray adaptation for the Nashville, Tennessee, squad car, shown above. Other changes were on the drawing board, and they propelled the company to the benchmark of the industry.

The all steel aerial, seen on this prototype for Lancaster, Ohio, remains conceptually unchanged through today. The tube steel frame under the skin of this sedan cab is also still in use today.

Anyone can be different; the key is to be innovative and create a basic long-lasting style. This is how Seagrave became the leader in the industry.

Quality

This view of the Arkansas city, Kansas, fire department is truly a picture of the past. So many early departments tried to create their own equipment on a passenger car chassis. These two Buick modifications served only a few years, but the 1918 Seagrave pumper and the 1931 Seagrave Quad each provided over thirty years of service.

Many cities, both large and small, learned the lesson of trying to modify or shop build. Some departments were successful, but most failed. How could a department hope to compete with a staff of engineers and designers whose only job was to create, test and build fire apparatus.

New York City ordered twelve ladder trucks from Seagrave, in 1925. Twelve more ladder trucks were ordered, in 1927, and then, in 1929, came a order for fifty hose wagons and a request to rebuild an earlier Hale water tower. Seagrave built their own spring raised aerial tower, but agreed to rebuild this water raised tower for such a good customer.

Water pressure provided the lifting force for this tower, through cylinders, sliding racks and the large quadrant gears. Five inch mains supplied the deck guns and a 3 1/2 inch hose supplied the duplex nozzles at the tip. The tower was reworked to a spring raised tower, in 1938, and served another 21 years.

This rig is unique because it was the only tractor trailer water tower made by Seagrave. The counter balance for the tower was provided by outriggers that folded against the frame when not in use. The poor tillerman hand to crouch beneath the mast, and his view was obscured by almost everything on the trailer.

The year, 1929, started out with a boom in New York. Mayor Walker's new fire commissioner, Jon Dorman, announced the starting of seven new engine houses as part of a planned forty new fire houses. The department put into service a new central alarm system, in Queens, and this 8,500 gpm water tower, rebuilt by Seagrave.

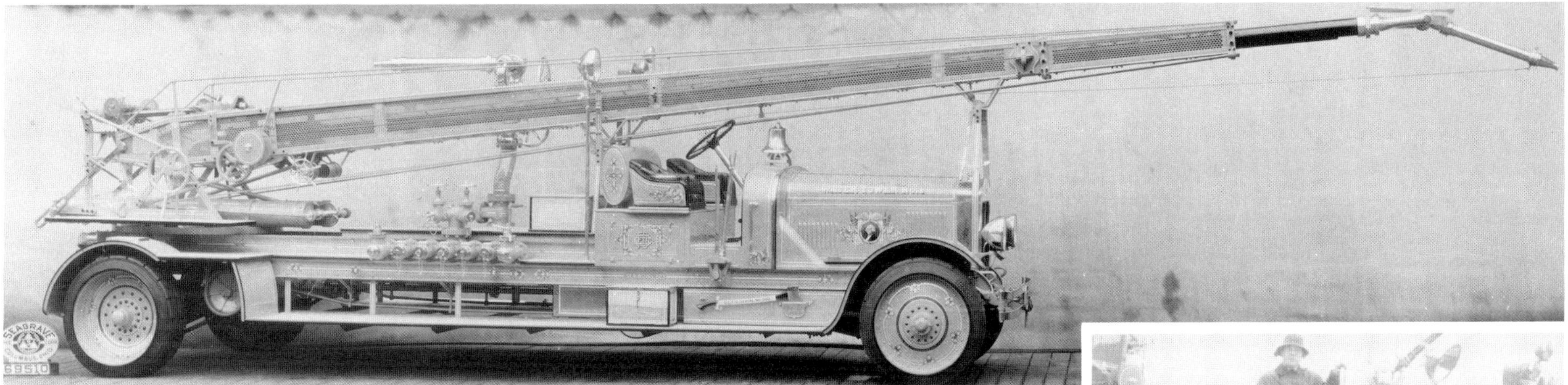

The last of the water towers by Seagrave was this 1932 delivery to our nation's capital. Its acceptance test brought out such dignitaries as Local fire buff, J. Edgar Hoover, in the trench coat. The tower was a spring raised 65 foot mast and a hand cranked tower apparatus. This delivery was the last of the chain drive Seagrave. The extra wide rear axle and deck gun were similar to the Detroit, Michigan delivery. The water tower's front end featured the round hood of a 1920's truck and the cowl of a 1930's rig (note the steering column on the outside of the hood). As metal aerial ladder rigs became common, the tower was raised less often. In August, 1956, the tower was scrapped..

Many dignitaries of the past were fire buffs. The name, fire buff, was coined by firemen as they noticed citizens huddling under buffalo robes at early fire scenes. Some of the more famous past butts were: New York City Mayor, Fiorello LaGuardia; First Lady, Eleanor Roosevelt; and Boston Pops conductor, Arthur Fiedler.

A true buff studies the fire scene the same way some folks critique a sports event. In many communities across the country, buffs run the fire department canteen, or participate in fund raising events.

Seagrave

2000 S. High Street
Columbus, Ohio

The plant on High Street is a collection of buildings dating back to 1895. Buildings one and two were built that year to house the Columbus Carriage Company. The buildings were of "brick and mill" construction, this is brick and timber. A brick exterior supported the outside edges of the beams, and wood columns supported the inside butt joints of the beams. In many areas the ceiling was 13 feet or less.

Building # 1 - built in 1895 - 11,000 sq. ft. on the first floor and 5,000 sq. ft. on the second. This structure housed the "Smith Shop" or rough assembly on the first floor and lumber storage above.

Building #2 - built in 1895 - 11,000 sq. ft. on the first floor and 8,000 sq. ft. on the second. Painting was done on the first floor. The second floor housed the wood pattern shop, saw sharpening shop, machine shop and dining room.

Building #3 - built in 1910 - Housed items from the old plant on Lane Avenue that was phased out earlier that year. First floor was used for final decorating, the second and third floors contained engineering and administration offices.

Building #4 - completed in 1910 - These original buildings formed a "U" shaped pattern. Its 35,000 sq. ft. were devoted to: first floor - erecting and testing; second floor - machine shop; and third floor - inspection and time study.

Building #5 - completed in 1928 - 23,500 sq. ft. total area. The first floor was a stock room and the second and third floors were pre-assembly and tool storage.

Building #6 - built in 1928 - Was devoted to sheet metal and frames on the first floor. The second floor was cab and aerial ladder development. Other out buildings, shown here, were the power house and foundry. Seagrave move the foundry into buildings across High Street, in 1952.

Today the plant is home to a number of sheet metal fabrication companies.

The Full Grille Era

1935 - 1951

On the shipping dock is a 1936 safety sedan pumper for Detroit, Michigan

The Full Grille Era could also be considered the era of the classic. From the mid 1930's, to the early 1950's, there was a refinement in style and performance. The fire apparatus matured in design and function. Positive features were enhanced and the result was a very reliable, fast and efficient apparatus.

On the style side, cabs were wider and more enclosed. The crew became sheltered from the elements. Tool boxes came off the running boards and folded into the basic design. Dual rear wheels became standard. The number of warning and work lights multiplied as traffic and firefighting became more complex.

As for performance, the V-12 engine would reign supreme. The big 903 V-12, first introduced in 1932, would, by the end of the decade, have a smaller partner. The Seagrave pumper would be one of the most powerful vehicles on the road, with almost twice the horsepower of most commercial trucks of the period. With the weight of the pump deep in the frame, the pumpers were remarkably stable. The traditional engine ahead-of-cab design, massive power and good weight distribution made these vehicles a pleasure to drive in heavy traffic.

There were a host of improvements worth noting that often go overlooked: hydraulic brakes, safety glass, sealed beams, rearview mirrors and heater/defrosters, just to name a few. Add the visible improvements to the hidden ones, such as, nylon cord tires, high octane gas and permanent anti-freeze, and the rigs of the late Full Grille Era are just as dependable as the apparatus in service today.

The top photos point out some of the subtle changes that took place in this era. The most often overlooked change was the width of the cab. This change took place after the war. Some more of the subtle changes took place in tire size, and the addition of sealed beam headlights. The federal government mandated a change to sealed beams, in 1940. The Seagrave, with the smaller V-12 and 6 volt system, made the deadline. The large V-12 powered vehicles were 12 volt equipped, and the changeover was slower due to lack of the non-standard (at the time) 12 volt sealed beam lights.

The photo, at left, was taken with the plant photographers own camera. The company used a large format camera with 11 x 14 inch glass slides for negatives, until 1920. The next camera featured a 7 x 11 inch gel negative. After the war, the negative size dropped to 5 x 7 inches. The format of choice dropped to 4 x 5 inches, during the 1960's. Photos today are shot on 120 film.

Every antique fire apparatus owner wants a factory photo. The truth is, not every delivery was photographed, and not every photo was shot under ideal conditions. The main reason to shoot early deliveries was to have a record of what was on the apparatus, when it was loaded in the rail car for delivery. Theft or damage during shipment could then be documented by the delivery photos.

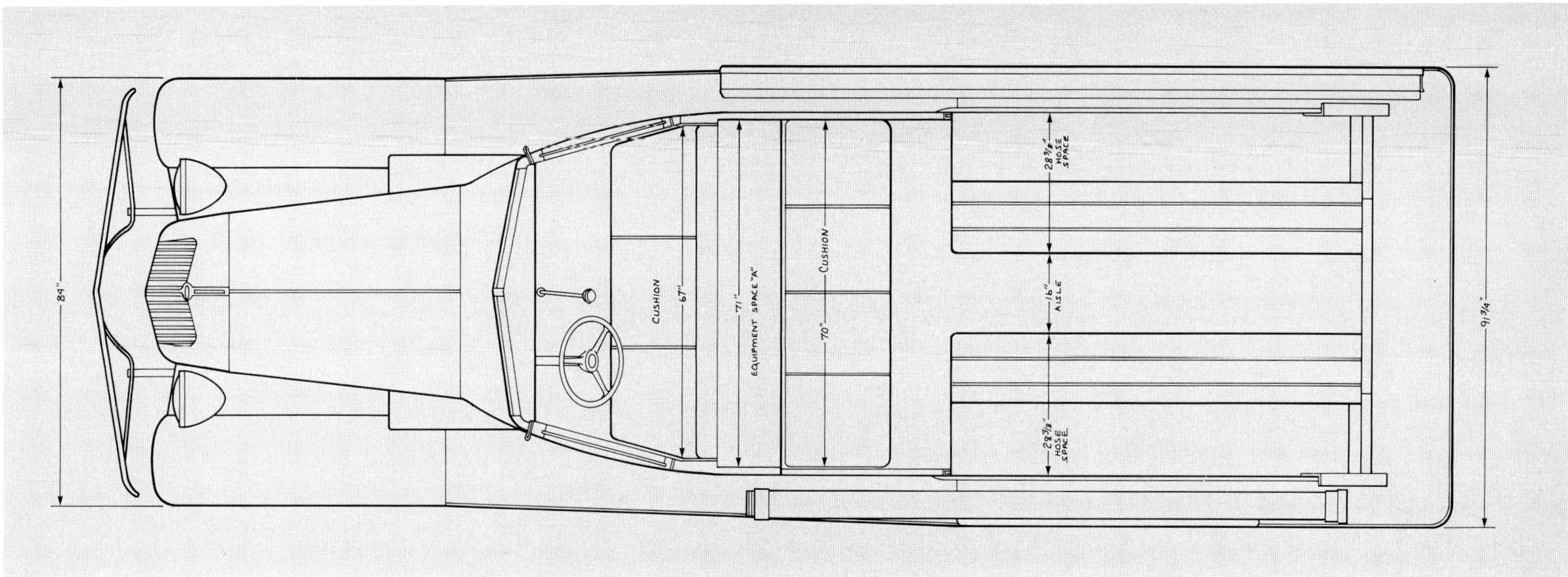
89"
CUSHION
67"
71"
EQUIPMENT SPACE "A"
CUSHION
70"
28 3/8" HOSE SPACE
16" AISLE
28 3/8" HOSE SPACE
91 3/4"

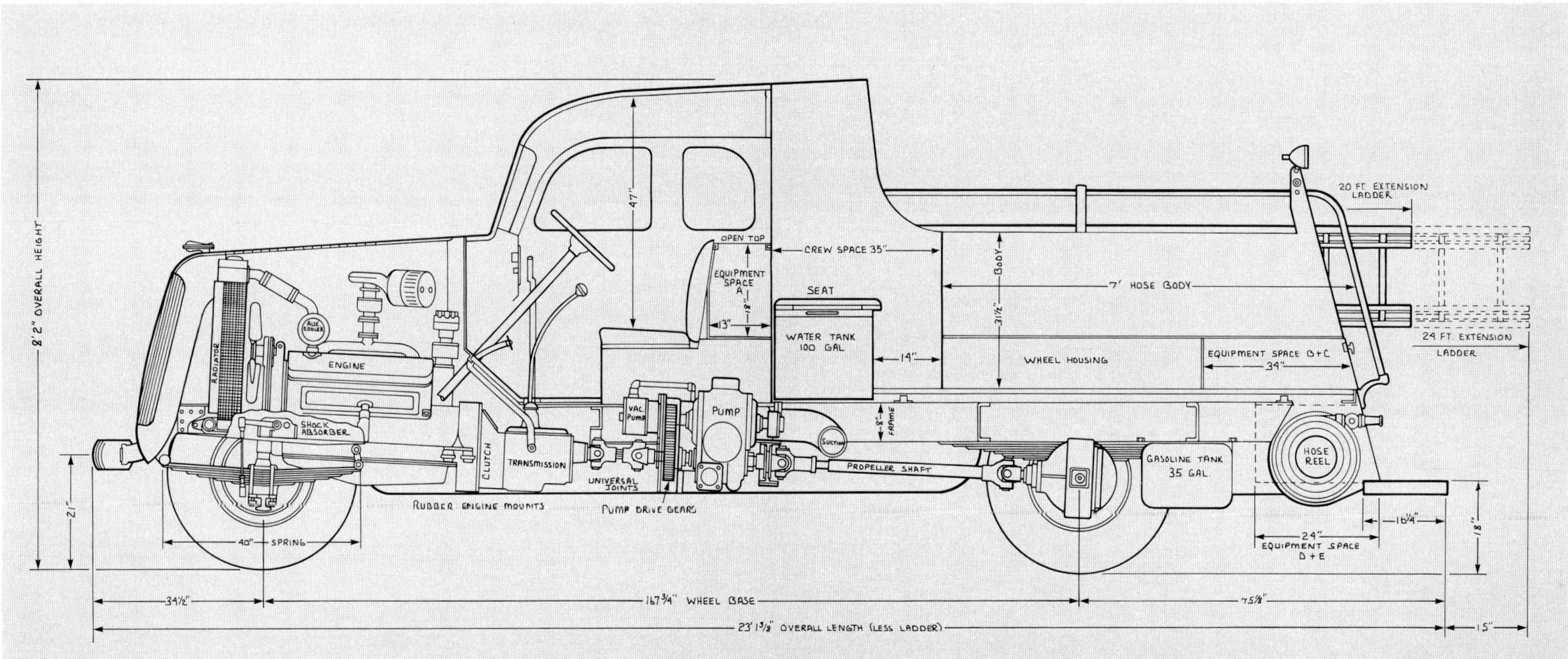
8' 2" OVERALL HEIGHT
47"
OPEN TOP
CREW SPACE 35"
EQUIPMENT SPACE A
18"
13"
SEAT
WATER TANK 100 GAL
14"
BODY
31 1/2"
7' HOSE BODY
WHEEL HOUSING
EQUIPMENT SPACE B+C
34"
20 FT. EXTENSION LADDER
24 FT. EXTENSION LADDER
RADIATOR
OIL COOLER
ENGINE
SHOCK ABSORBER
CLUTCH
TRANSMISSION
UNIVERSAL JOINTS
VAC. PUMP
PUMP
8" FRAME
SUCTION
PROPELLER SHAFT
GASOLINE TANK 35 GAL.
HOSE REEL
RUBBER ENGINE MOUNTS
PUMP DRIVE GEARS
21"
40" SPRING
24"
EQUIPMENT SPACE D+E
16 1/4"
18"
34 1/2"
167 3/4" WHEEL BASE
75 5/8"
23' 1 3/8" OVERALL LENGTH (LESS LADDER)
15"

SEAGRAVE CANOPY CAB

The Seagrave Canopy Cab

This is probably the most distinctive and fashionable style to ever flatter a fire engine. The production run of Canopy Cabs lasted until 1970. They were built for fourteen years, in the Full Grille Era, as well as, nineteen more years during the anniversary series.

Rumors in the industry, in the early 1930's, declared Mack and others were working on an enclosed fire engine. Not to be outdone, but not in a race to be first, the Seagrave engineers set out in the traditional sheet metal over wood construction. This style was used in the enclosed box shaped cars of the day. In 1936, Seagrave built a few wood and sheet metal sedan cabs for Detroit. During assembly of these units, it became apparent that a structure, the size of an enclosed fire truck, must be all steel and well reinforced.

What evolved from the prototype enclosed cabs was the canopy cab style. It was an arrangement where the driver and officer would sit back to back with the crew. The back portion of the cab, the upright with the rear window, became sort of an internal roll bar. True, some hose box space was lost to the walkway, but now the whole crew could be seated and safe.

This layout could accommodate seven men, two compartments for 750 feet of hose each and a 100 gallon water tank.

The vehicles shown here are: the first canopy cab for Bar Harbor, Maine, a 1936 delivery; a war time 1942 pumper for Wilmette, Illinois; a post-war 1949 delivery to New London, Connecticut; and a more recent photo of a rig for North Bridge, Massachusetts. Even the addition of modern lights and a radio antenna cannot detract from the classic design.

PORT CHESTER FIRE DEPT.
SEAGRAVE
THE CORP
COLUMBUS, OHIO
80895

S.F.D.
SEAGRAVE
COLUMBUS, OHIO
A2430

Canopy cabs were only part of the product mix during the Full Grille Era. Many fire departments preferred open cabs. The deliveries shown here are for Port Chester, New York, 1936; Springfield, Massachusetts, 1939; Mount Vernon, New York, 1942 and Rothschild, Wisconsin, 1948.

Note how the driver became more enclosed, culminating with the addition of the side glass. The angle of the windshield changed too, after the war. Inside the cab, the dash was also placed on a slight angle so that glare might be reduced.

The grille shells of this era came in two basic designs, the sweetheart grille and the waterfall style grille. The difference in grille shape and hood configuration are depicted in the photos, at the left.

Smaller communities like Pembroke, New Hampshire, and Sherman, Texas, chose smaller cabs. The 1940 delivery to Pembroke featured a Highland two-man cab.

Highland Body Company of Cincinnati, Ohio, built custom truck cabs for some Seagrave, Ahrens-Fox and American LaFrance fire apparatus, over the years. They also built bodies and cabs for Railway Express and Caterpillar Tractor.

The Highland Buggy Company was formed in 1870, as a custom buggy builder. The company changed its name to The Highland Body Company, in 1920. During the depression, it became the Highland Division of the Trailmobile Company, a large truck trailer manufacturer. In 1948, the cab building business was sold off to become The Truck Cab Manufacturing Company, still of Cincinnati. Many custom apparatus today use the Cincinnati cab. Although Seagrave builds its own cab today, some of its cab hardware comes from this company with a history as long as Seagrave.

In 1939, Seagrave delivered this four door, closed cab, pumper thirty years ahead of the trend to enclose pumpers. The apparatus was streamlined; even the suction hose and ladders were enclosed.

The seating arrangement would accommodate eight persons. The driver and officer would share the front seat. Six firemen could sit on side facing bench seats. Everyone was protected from the elements.

Engine 1 ran with Ladder 1, also a four door enclosed cab. The last run of Engine 1 was on May 12, 1951. While responding to an alarm, Ladder 1 hit Engine 1 at 30th and Evelin Streets. Eleven fire fighters were injured. (Ah, the days of large crews.) Thanks to the enclosed cab, no one was killed.

Engine 1 was damaged beyond repair and was scrapped. The trailer of Ladder 1 was fitted with a new Seagrave tractor and answered alarms for many more years. The new Engine 1 would be a 1951 Seagrave 1,250 gpm pumper.

The photo, above, shows a 1936 Detroit, Michigan, sedan after an accident with an Ahrens-Fox fire high pressure hose truck. Both rigs were answering the same alarm on August 20, 1936. The sedan was only months old. The hose wagon hit this Seagrave on the side, enough to bend the frame seven inches, in the middle (note the rear wheel location in the fender opening). No men in the Seagrave sedan were injured, but both men riding the Ahrens-Fox were seriously injured. The wrecked sedan was rebuilt and placed back in service. Analyzing the results of the accident on structural body members led to the development of the steel sub frame, shown at the left. Three more sedan style pumpers were delivered to Detroit, Michigan, in 1936, four in 1937 and five in 1938.

The 1938 deliveries boasted a steel body structure. The earlier deliveries were a wooden frame. This steel skin over a tube steel frame provides maximum support with a minimum of bulk. This same design is used in Seagrave cabs today, fifty years a proven design.

The sedan design, of 1936, brought all fire fighting equipment under cover. Suction hoses and ladders were even inside. The crew sat on a bench seat in the middle of the apparatus facing the curb side.

Engine 3 was a 750 gpm pumper, of the new streamlined style, delivered in 1938. It was on display at the Fire Chief's Convention in New Orleans, Louisiana, and shipped by rail to Wakefield, Massachusetts, after the convention. The fire chief decided on this style pumper after viewing the 1937 sedan of Gorman, Maine. The rig cost $9,500, weighed 12,200 pounds and was in front line service for 22 years. Wakefield, Massachusetts, liked the sedan style so much they copied it on a 1941 Dodge chassis.

Shown, at right, is the 1941 Dodge sedan cab answering to an accident involving the city's 1928 Seagrave ladder truck.

When we speak of the Los Angeles Fire Department, we must specify L. A. City or L. A. County. They are different fire districts. The county had responsibility for the large semi-rural area. The city department covered dwellings through the largest high rises.

Delivery #88040 was a 1938, special manifold wagon, for L. A. City. This rig served as part of a two part engine company, Engine 3. Besides a small tank and booster pump, this truck carried 2,000 feet of 3 1/2 inch hose at the rear and 1,500 feet of 2 1/2 inch hose in the cross bed compartments. The turret gun had up to 3 inch tips.

Los Angeles County was proud of its 1938 pump and hose wagon, nicknamed the "Galloping Ghost". This truck weighed 52,000 pounds when loaded with 2,500 gallons of water. It also carried 1,200 feet of 2 1/2 inch hose and 1,350 feet of 1 1/2 inch hose. Assigned to the Norwalk district of L. A. County, it originally covered sixteen square miles. Moving this big machine around that area for years caused a nickname change of the rig to "Big Bertha".

L. A. County's pumpers often carried large water tanks like the delivery to Engine 40. Some of L. A. County's apparatus were painted green with a single black stripe like the classic delivery for Engine 8.

L.A. COUNTY
FIRE DEPT.
SEAGRAVE
COLUMBUS, OHIO
91005

FIRE DEPT.
ENGINE 40
SEAGRAVE
THE
CORP
COLUMBUS, OHIO
90465

SEAGRAVE IMPROVED CANOPY CAB

At first glance the two deliveries, shown left, one to Columbus, Ohio, and the other to Baltimore County, appear identical. But the difference lies in the series, Standard versus Sentry. Note the hood length, cab style, fender opening, width of cowl, grille shell, bumper and wheel type. Those are just a few of the changes all the way back to the flair of the body as it meets the tail board.

By the 1940's, decorator's art had been mostly relegated to the apparatus doors. Shown are:

Tarrytown, New York	Freeport, New York	Winchester, Virginia
Ypsilanti, Michigan	Alliance, Ohio	Elgin, Illinois
Greenwich, Connecticut	Middletown, New York	Lansing, Michigan

The change in pump panels took place after the war, but it would be easy for an engineer to adapt to these simple controls. Studying the upper photos, you will notice the reflection of the photographer in the shiny suction cap.

Seagrave ran their own chrome shop. Steel items were triple chrome plated: that is, copper plating, then nickel plating, then chrome plating. Even brass bells were copper plated first. The copper plating enhanced the tone quality of the bell.

The pump throughout the period was an offshoot of the Manistee style B pump, introduced in 1920. The designation now, however, was just the B pump. All pumps were made in Seagrave's own pump shop.

The difference of the pump panels is as plain as the difference in colors of these two engines. The deliveries to Pennsylvania ran from yellow to green and almost everything in between. The Bellevue, Pennsylvania, delivery was dark maroon. Shippensburg, Pennsylvania, was light brown and light red.

No. 1
BELLEVUE
BELLEVUE
SEAGRAVE
A5985

VIGILANT HOSE CO.
No.1
SHIPPENSBURG
PA.
COMMUNITY
EQUIPMENT
F-680
THE

One of the greatest developments, of the 1930's, was the all steel, hydraulic and aerial ladder. Wood ladders were serviceable, and many communities requested wood aerial ladders well into the 1940's. But wood ladders required maintenance and were very limited as to weight and stress. The wood would swell and twist, making extension of the fly sections difficult. The all welded steel aerial ladder, a Seagrave first, eliminated the faults of the wood ladder.

Shown here on a round hood chassis is the prototype ladder. Through testing, the base of the ladder was reinforced, and the height of the truss/hand rail was raised.

On November 15, 1937, thirty-five Seagrave employees stood on the base section of an 85 foot aerial ladder. The total weight of the men was 5,822 pounds. The total deflection of the ladder was 5/8 of an inch. This ladder follows basic truss design of the very first Seagrave ladders of fifty years previous. The design is still in use today, fifty years later.

During the Depression years, Seagrave was busy on the new designs. The photos, above, show the interior of the seven man cab for Detroit. The crew sat on a bench seat across the back of the cab, and the jump seats attached to the back of the driver's and officer's bucket seats. This new design cab and 100 foot aerial trailer, shown here in test, were delivered to Detroit, Michigan, in 1937. Optional seating arrangements were the open cab or semi-canopy cab.

The seven man cabs were an innovation that is in a revival today. The base price for an 85 foot tractor-drawn aerial truck then was $18,500. The canopy cab was a $500 option, and the four door cab was a $1,500 option. Shown are rigs for Norristown, Pennsylvania; Cleveland, Ohio; and Detroit, Michigan. A 75 foot aerial was $250 less, and a 100 foot aerial was $500 more.

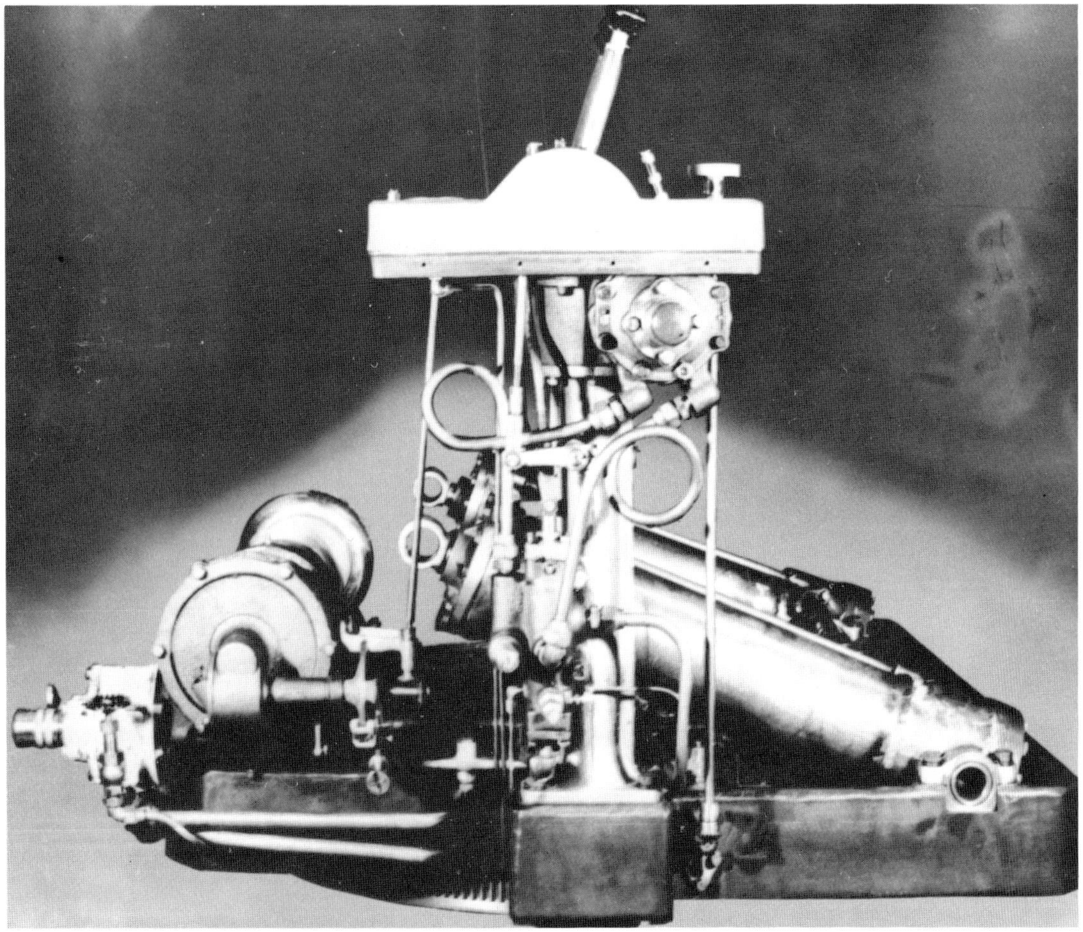

The hydraulic lift mechanism has remained basically the same, since 1935. A hydraulic tank in the frame supplies fluid to the transmission power take off (PTO) driven pump. The fluid then either fills the lifting cylinders or turns the winch drum. Only the cylinders require large amounts of liquid. Fluid to turn the drum is immediately returned to the hydraulic tank.

At first glance, these aerial trucks look identical, but there are many differences. The Washington, D. C. truck was built in May, 1940, and the Cincinnati, Ohio, truck in August. Both are the large 906 cubic inch displacement engine powered 100 foot aerials. Differences included: sirens, Roto Rays, headlights, lanterns, compartments, and more.

To tiller one of these rigs is an experience. You're almost ten feet from the ground. Traffic is gawking at you and not really getting out of the way. You have to coordinate with the driver to maneuver this up to 15 ton truck, and you have no brake pedal. As the photo shows, there isn't much room . . . even less with turn-out gear.

The fellow in the tiller seat is the same one who stood on the idle of the ladder, on page 143, now a few years older and maybe a little heavier.

During the 1940's, the debate on ladder design would rage on. The diehards still favored wood aerials. New York City ordered seven wood 75 foot aerials, in 1941.

Some competitors used a riveted aluminium construction. Early aluminum aerials had tendencies to deflect and take a heatset. Part of this was due to the fact that they were of riveted construction. The process of welding aluminum effectively is relatively new. If the early aluminum ladders got too close to the fire (above 600°F), they lost over half of their strength.

Seagrave ladders were all steel. They were fully welded making the bolts and rivets of competing makers unnecessary.

The Seagrave truss wood ladder was the standard of the world, since 1881. The all metal Seagrave aerial ladder combined the proven aspects of a truss design with the solid dependability of steel.

When comparing the photos, on this page, note the overhang of the 100 foot aerial for Manchester, New Hampshire, to the 85 foot aerial for Chattanooga, Tennessee.

The two aerials are from each end of the country and span the World War II period. Serial #A2510 is a 1939 delivery for Portland, Oregon; and serial #D1610 is a 1948 delivery for New York. This was New York City's first Seagrave metal aerial ladder truck. The previous 52 aerial ladder deliveries had all been wood aerials.

Portland, Oregon, had the only four door open cab ever built. These trucks were typically too long and heavy to photograph on the loading dock, so they were driven to the other side of the railroad viaduct from the plant. The house that appears in both of these photos can also be seen on page 204, at the top of the plant layout photo.

These pages depict a first and a last. Edwardsville, Illinois, received the last Seagrave ladder truck powered by the F-6 engine. From 1915 to 1936, Seagrave F engines powered the majority of Seagrave ladder trucks.

In August of 1936, the same month as the delivery to Edwardsville, Seagrave introduced this new design, for Grandview Heights, Ohio. This delivery featured: a new style grille, a new Pierce-Arrow built V-12, and the first use of preconnected crosslays.

Grandview was a suburb of Columbus, Ohio, and Seagrave's plant manager lived across the street from this fire station. Consequently, this vehicle was shown to many prospective customers from far and wide.

The addition of a volume pump to a ladder truck, of any type, made an almost complete fire department on wheels. What would the wagon masters of only thirty years previous think of these rolling fire stations?

The 1940 prices for a Quad (ladders, hose, water tank, and volume pump) started at $10,900. The price for a Quint (Quad plus aerial ladder) started at $15,150. Prices were for three man open cabs with small V-12 engines. A closed cab was a $500 option and the large V-12 was a $2,000 option.

The apparatus shown here were a quad for Jericho, New York, and a quint for Tuscaloosa, Alabama.

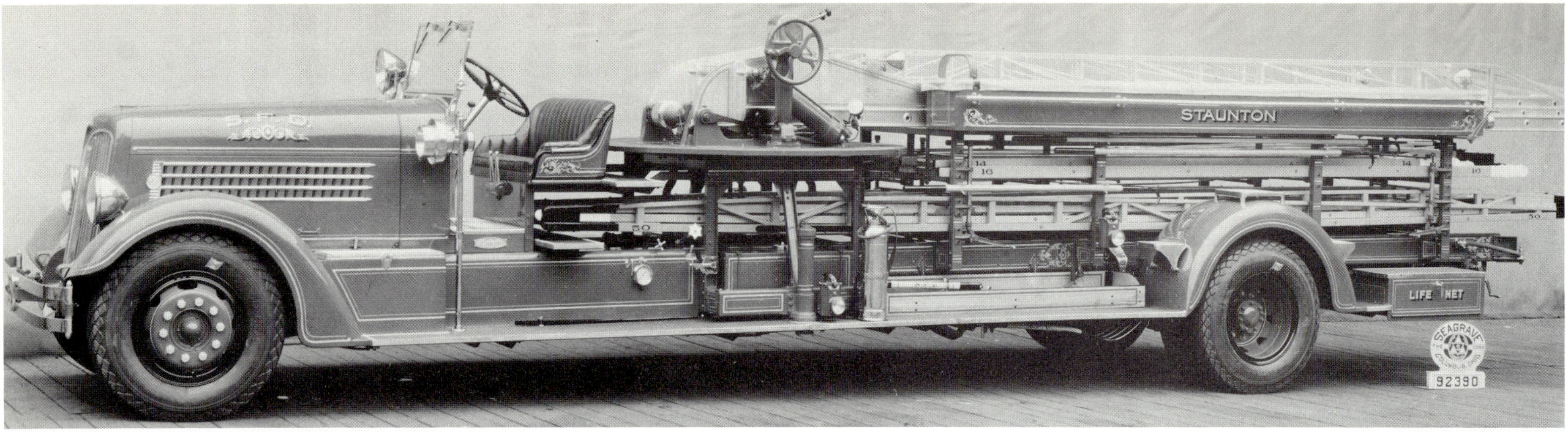
STAUNTON
LIFE NET
SEAGRAVE
92390

LADDER No. 1
CONCORD
LIFE NET
SEAGRAVE
A5415

LADDER NO. 2
LIFE NET
SEAGRAVE
THE CORP
COLUMBUS, OHIO
A8225

Close inspection of the photos on these two pages show the evolution of the service aerial truck. The first delivery shown is to Staunton, Virginia - 1938. Subsequent photos are: Concord, New Hampshire - 1941; Muskegon, Michigan - 1941; Memphis, Tennessee - 1944; and Saginaw, Michigan - 1946.

The enclosed body compartments and improved seven man seating greatly enhanced the functional use of these rigs. All of these trucks feature the standard three section, 65 foot aerial ladder. There were a few four section, 85 foot service aerials built, starting in 1941 (see page 238).

In the pre-war deliveries, the length of the hood, and its relation to the rear edge of the fender, gave some indication as to the power plant under the hood. The big V-12 was ten inches longer than the small V-12. With the introduction of a new cab style, after World War II, Seagrave tried to standardize parts and designs. So, on the big V-12, the radiator was moved forward 10 inches, the hood was longer and the front fenders became these large clamshell looking affairs.

Post-war city service aerial ladder trucks were 41 feet long, 8 feet 7 inches high, 8 feet wide, and had a wheel base of 22 feet. They weighed 22,500 pounds and cost a little over a dollar a pound.

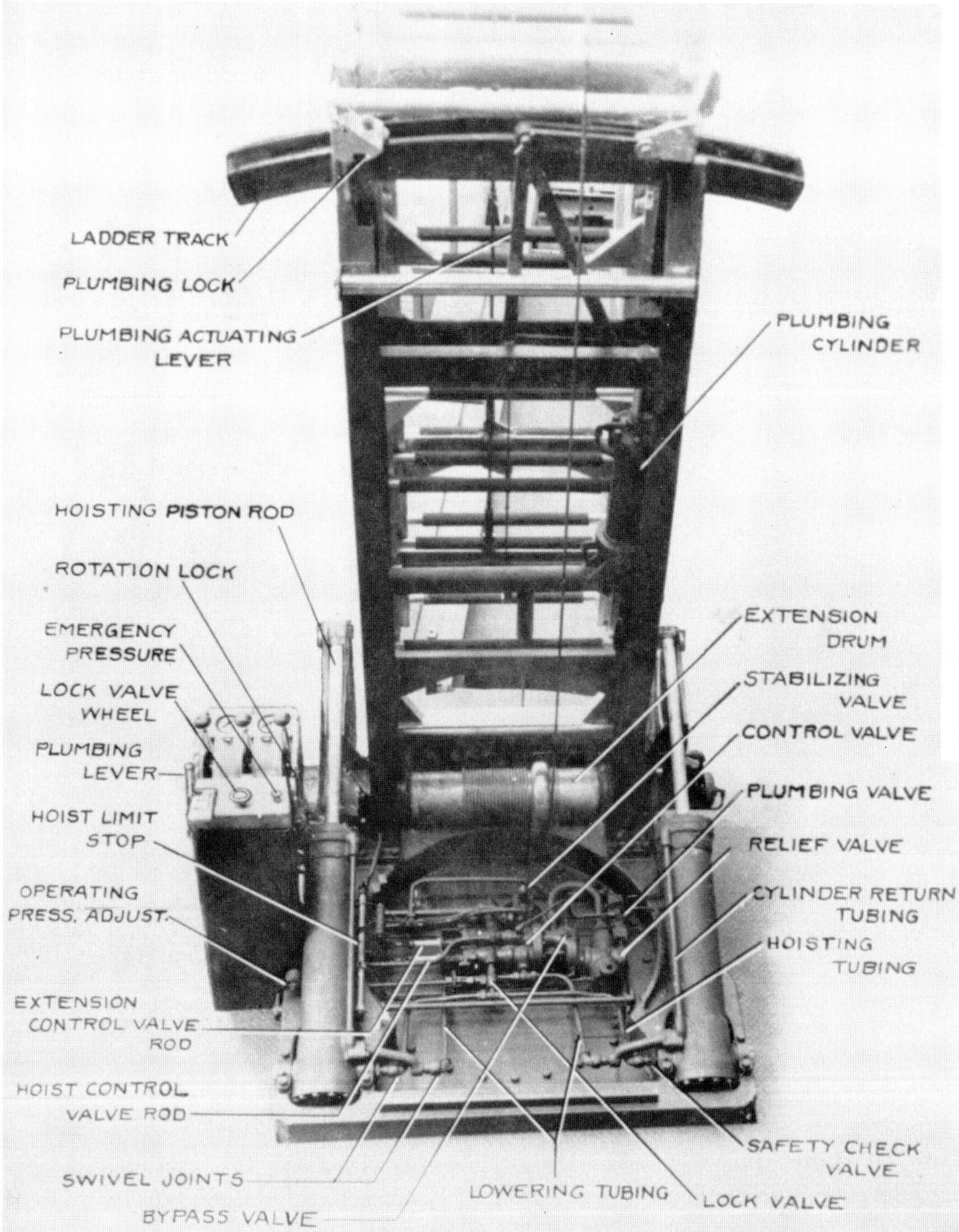

The photo, at left, shows the ladder controls for the pivoting style ladder trucks. If you parked your ladder truck on a hill, it would be difficult to raise the aerial ladder vertically square with the building. The theory for this ladder was that the base section would pivot, at its base, and ride on a track to guide the ladder 10°, or so, from vertical.

Plumbing of the ladder was controlled by linkage and a hydraulic cylinder, shown under the pivot base. A side view of this ladder showing the enlarged base section is on page 246.

The Magirus aerial was popular in Europe. The multi section ladder was available in sizes over 100 feet. Cities desiring longer ladders often asked for this style ladder, but very few departments ever reordered them. The Magirus ladder, in the photos above, went to Montreal, Quebec, in 1949.

The novel ladder rack arrangement for Albany, California, proved too cumbersome. Each time you had to reload hose, you had to swing away the ladders.

The penalty of leadership

To remain a leader in the first apparatus business, Seagrave has always tried to exceed its customer's expectations. Some new items were successes, some were not. For example, when the market was happy with piston and gear type pumps, Seagrave introduced the centrifugal pumps.

The Seagrave firsts are impressive. Even if someone else beat Seagrave by a few months, the best in class awards are more coveted. The Seagrave firsts include:

- 1912 - Centrifugal pump & pressure regulator
- 1915 - Auxiliary cooler
- 1935 - All steel hydraulic aerial ladder
- 1937 - All steel safety cab
- 1938 - Positive safety ladder locks

But there were a few non-successes: pivoting aerial ladders, Magirus style rear mounted ladders and fold-away ladder racks.

SEAGRAVE
THE CORP
COLUMBUS, OHIO
A1065

W-50805
SEAGRAVE
THE CORP
COLUMBUS, OHIO
A9165

The government started to reactivate and update military bases, with the threat of war looming, in the late 1930's. Through the war years, Seagrave produced hundreds of standard 750 gpm pumpers, for the U. S. government, often in batch orders of fifty or more.

Delivery records indicate that within two weeks after the bombing of Pearl Harbor, apparatus deliveries, to some communities, were being diverted to military installations. In the first year after Pearl Harbor, the government ordered hundreds of apparatus for defense plants and military destinations.

It's interesting to note the disappearance of chrome and trim on these wartime deliveries. After the war, Seagrave sent letters to communities indicating that some of the painted trim pieces were actually stainless or chrome under the paint. Seagrave had stocks of shiny parts on hand before the war began and painted over the gleam to meet government mandates.

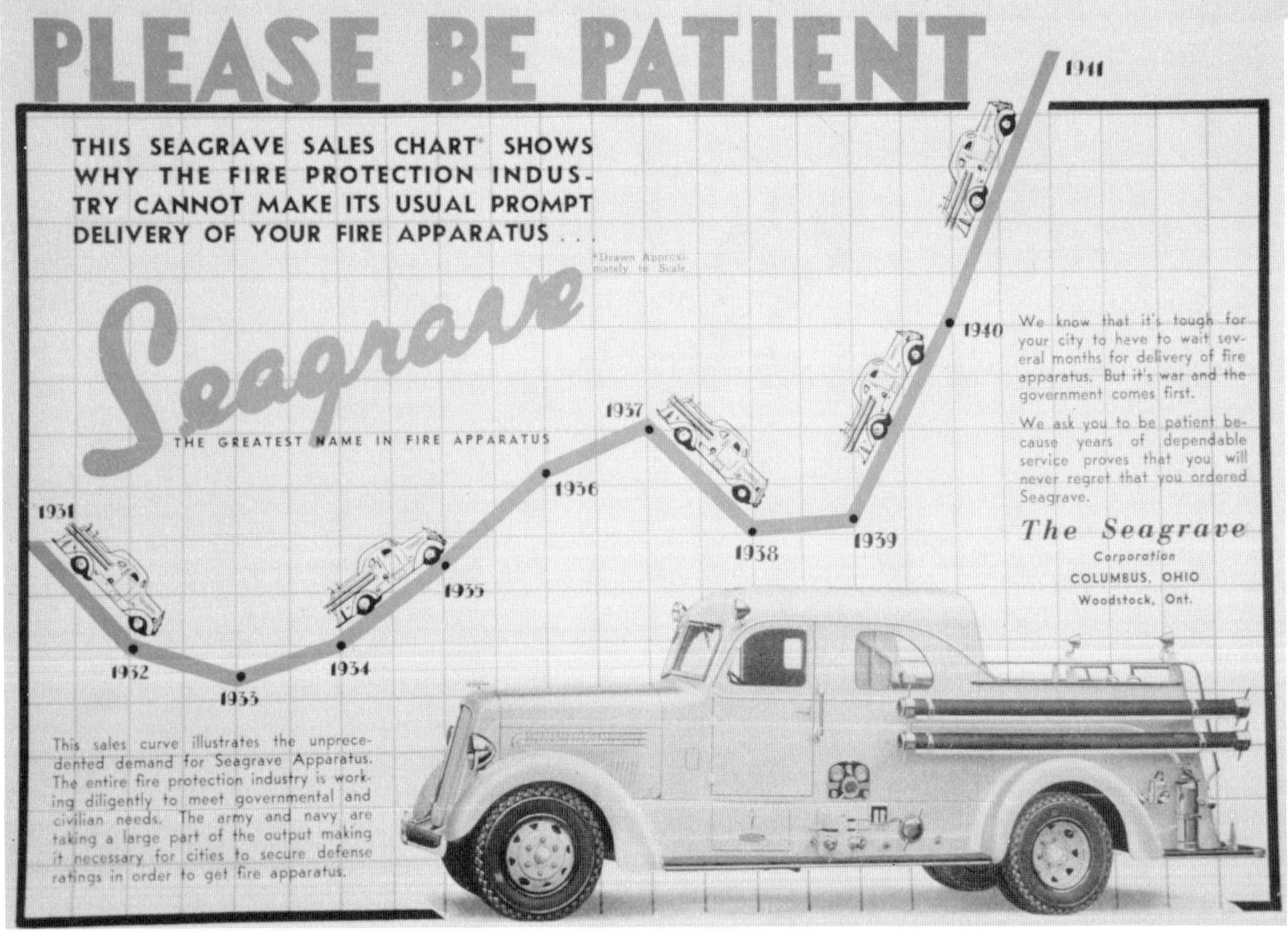

Seagrave built four, 85 foot, four section service aerial ladder trucks to test this new design. One went to St. Johns, New Brunswick, one to Great Falls, Montana, one to Pittsburgh, Pennsylvania. The fourth build was to be the factory demonstrator. But when the Navy needed replacements for equipment lost at Pearl Harbor, who could say no.

The government mass orders subsided with the turning point in the war. And by late 1944, the delivery list contained only industrial users and municipalities.

Seagrave equipment appeared on some commercial chassis, for military use, as well. International chassis were delivered as both standard pumping units or chemical trucks. Seagrave booster pumps and tanks were often attached to big Studebaker 6 x 6 trucks.

Many of the wartime deliveries stayed overseas, as did many of the deliveries of the first world war. One story is often told of a U. S. soldier, in France, in World War II seeing a 1917 Seagrave with his hometown's name still on the hood. It was one of the preempted deliveries from World War I.

During the war, one of the key slogans was "Loose Lips Sink Ships". Even at Seagrave, the delivery destinations of war time deliveries were shrouded of mystique. The Army deliveries in the paint log listed one serial number with the note "50 machines". Final destination of the apparatus was kept secure so no one could spot troop or base movements by looking at records. What can be obtained from the records is the fact that the government purchased over 125 - 500 gpm pumpers on International chassis, in 1943.

SEAGRAVE
THE CORP
COLUMBUS, OHIO
87610

No. 3

Seagrave built on whatever chassis the buyer specified. Ford and International chassis were favored because of horsepower and torque capacities. Seagrave equipment on a Diamond-T chassis, as shown above, was a rare item. The cost of the expensive Diamond-T chassis and equipment would have been close to that of a full custom Seagrave.

The majority of commercial deliveries were to industrial companies or small towns. A Seagrave commercial delivery could give a community the dependability of the Seagrave pump with the driveline a "shade tree" mechanic could handle.

Deliveries shown here are for: Eufaula, Oklahoma - 1937; St. Matthew, Kentucky - 1941; New Lexington, Ohio - 1939; and Wakefield, Massachusetts - 1950-53.

The 1950 Wakefield Dodge hit Wakefield's 1925 Seagrave aerial ladder truck, in 1953 (see page 215). It was sent back to the factory and the fire equipment was mounted on the 1953 Dodge chassis, shown here.

Seagrave delivered 210 pieces of apparatus in 1950, about one for each working day of the year. These deliveries included sixteen commercials: four Chevrolets, one Dodge, seven Fords, and four Internationals.

The Frostproof, Florida, Fire Department took delivery of their new Ford, single stage, 500 gpm pumper in December of 1947.

This was the new look for Ford for the 1948 and 1951 model years. The Seagrave pump on the other hand was virtually the same since 1932.

Commercial trucks were a necessary part of the business, especially if cost was the major factor. But, beyond cost, the differences were staggering. The Ford's frame was 1/4 inch thick, the Seagrave frame was 5/16 inch thick. Ford's "Rouge 6", the biggest engine for this F-6 chassis, had 122 horsepower. Seagrave's smallest engine, the old straight 8, was rated at 150 horsepower.

Usually an upgrade to a custom Seagrave was an easy sell, unless overall budget was a factor. The average commercial chassis of this period even looked Spartan in appearance, when compared to the custom deliveries.

Over the years, Seagrave supplied pumps and chassis to a few selected builders. Howard Cooper Corporation of Portland, Oregon, purchased Seagrave pumps from 1936 to 1946, for mounting in commercial chassis.

Yankee Walter of Los Angeles, California, would buy components and build the custom cabinet work. The most interesting Yankee Walter delivery is the Seagrave aerial ladder (with bucket). It's mounted on an International KB-8 chassis with a Coleman four wheel drive conversion.

The plant on High Street made more components for its vehicles than anyone since Ford and his Model 'T'. Ladders, hoods, frames, grilles, pumps, and engines were just the start. Just think of all those custom castings. Employment in 1940, the year these photos were taken, rebounded to the pre-depression level of 350. Seagrave stock was now trading at four versus the 1932 low of one.

Plymouth, Massachusetts, ordered this aerial ladder truck in June of 1940. They sold it in April, 1991. During its fifty plus years of service, the outside remained virtually unchanged. This six-cylinder Seagrave G engine was replaced after 28 years of service with a GMC V-6. The truck was still operational when sold.

The base price of a 65 foot open cab, six-cylinder Seagrave, in 1940, would have been $11,900. Standard items were booster pump, hose tray, 100 gallon water tank, and 240 feet of ground ladders.

Options were: rear step - $100; seven-man cab - $500.

Credit could be had for deleting: ladders - $600; hose tray - $250; water tank - $250.

The company allowed a 5% discount for cash. Apparatus financed through the company carried a 6% interest rate. Trade-in allowances were: 70% of original price the first year, 55% the second year, down to 10% the thirteenth year.

The folks at Three Rivers, Quebec, ordered this 100 foot three section aerial on October 14, 1936. It was the first custom cab Bickle Seagrave made. It left for the Bickle Seagrave factory in February, 1937. Export versions to Canada were subject to heavy customs duty. Add this tax to Canadian loyalty for Canadian built products and you have good reason to ship kits and not complete apparatus to Canada.

Quebec City, Quebec, ordered their first Seagrave, in 1911. The subsequent orders continued in 1923, 1931 and 1935. There were six deliveries during the full grille era. The narrow-bodied, 1944, 85-foot aerial is shown center. All of the others are wide body post-war deliveries.

The unit, above, went to Lachine, Quebec, in 1944. The semi-completed rig would be finished, painted and trimmed out in the Woodstock, Ontario, plant - right down to the radiator emblem shown at the left. Bickle Seagrave built 203 custom deliveries between 1937 and 1956.

The photo, right, shows a close-up of the fifth wheel, trailer hitch mechanism. The trailer is permanently attached to the tractor by inserting two pins through the trailer gear into the eyes cast into the fifth wheel casting.

When a 100 foot aerial ladder is raised and swung sideways, the load is carried by the inboard spring. To alleviate this load, spring locks are engaged. The locks, shown right, mechanically connect the frame and the axle. They are engaged by pulling the large handle located on the outside of the cab, to the rear of the driver's door.

The locks are actually two serrated disks that are forced or locked together. This action prevents a weight shift and consequent spring compression or sag.

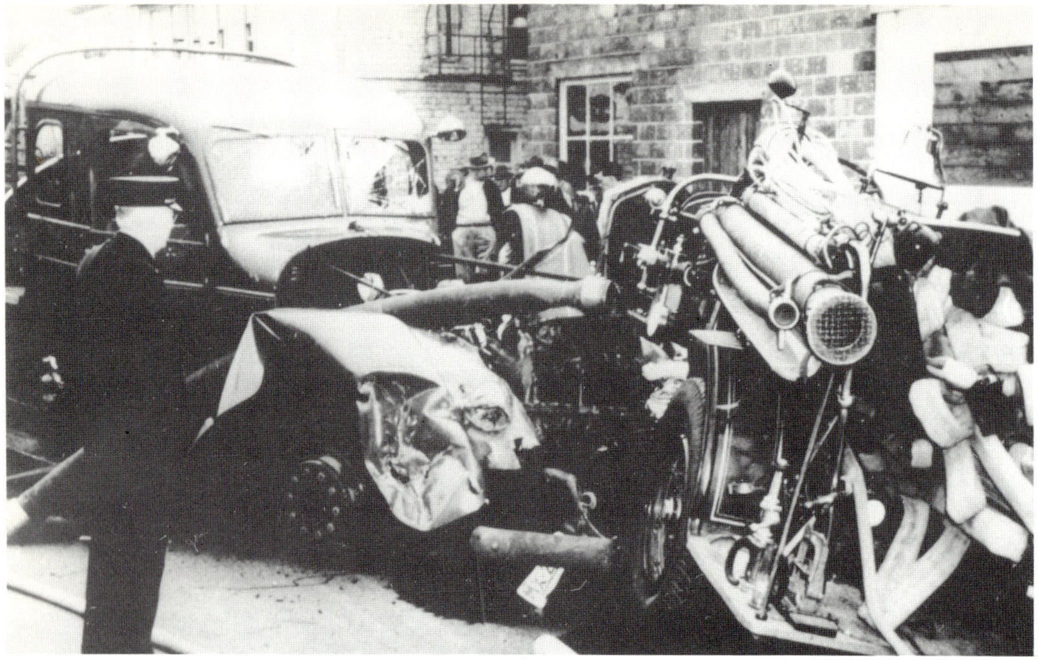

Fire apparatus is a true life saving piece of equipment and, therefore, deserves the right-of-way in traffic. The average fire doubles in size every minute, in its early stages. Immediate access to the fire is imperative to save lives, as well as, property.

The accident photo, top left, was taken at Aberdeen, Washington. The accident, above, happened in Columbus, Ohio. The photo, at left, shows a 1941 delivery to Harrison, Virginia, back at the factory for a rebuild after an accident, in 1955.

Seagrave has always rebuilt apparatus for valued customers. Who better to rebuild than the original manufacturer? Page 70 shows the first total rebuild. Even very resourceful Detroit shops sent their new sedan cab back to the factory (page 212) when they learned the frame was bent inward seven inches.

Pittsburgh, Pennsylvania, sent this 1941 service aerial back to the plant for repairs, in 1945. The rework took place in what the employees referred to as the "Blacksmith Shop". Here, old time smithies were to fabricate and forge a rig back to new condition. This truck is having repairs made to its air over mechanical brake system.

This photo, right, of the Columbus Fire Department shops must been typical of scenarios played around the country.

"How did this happen . . . I don't know."

In many communities, Seagrave is a tradition

The photo above shows part of the New Haven, Connecticut, fleet in one of the saddest occasions, a funeral procession.

In stark contrast to the all white deliveries of New Haven, were the deep red colored deliveries to Methuen, Massachusetts.

Classic in style from the first to the last

Detroit, Michigan, was a large Seagrave user for years. The Detroit apparatus garage looked like a Seagrave facility. In fact, Detroit had so many Seagraves, they hired a repairman away from the Columbus factory and made him shop superintendent.

Having your whole fleet from one manufacturer makes sense. As older rigs retire or are damaged, their parts become spares for the current fleet. In hard urban service, like Detroit, the average life of a Seagrave was always greater than the life span of commercial deliveries for neighboring departments.

The photos show distinct differences in the two Seagrave grilles of this era. The ladder truck was a 1950 delivery with a waterfall grille. The 1936 pumper has what is commonly referred to as a sweetheart grille.

SEAGRAVE
SEAGRAVE
SEAGRAVE
SEAGRAVE

The big V-12 story

During the late 1920's, the world seemed to be in a multi-cylinder race. Motor boats, airlines, zeppelins, and race cars were all powered with V-12 engines. Rival American LaFrance was rumored to be ready to scoop the fire truck field, with a V-12. Seagrave engineers also toiled with their version of the V-12 engine.

Although the rival V-12 was out first, it received some bad press when it failed to appear at some early demonstrations.

The Seagrave 906 cu. in. V-12 engine, introduced in 1932, was designed for fire service and easy maintenance. Dual ignition was designed in, not added on, like some commercial engines, and the valves were even replaceable without removing the head. The ports to remove the valves can be seen just beneath the spark plugs. This feature was very important because gasoline quality was not stable in this era, and carbon build up on the valves of gasoline engines was common. The carbon would cause the valve not to seat properly, causing poor performance.

The inset of the first design, and the full photo of the 1940's version, show slight changes in ignition and carburetion, which raised the horsepower from 240 to 268. The early V-12 engine had Stromberg DDR-5 dual throat down draft carburetors. Later versions had dual Zenith carburetors. The big V-12 was test run on a dyno-tester, in the plant for 24 hours, before installation in the chassis. The statistics of this engine reflected its size. For example, it weighed 2,350 pounds. It took 38 feet of wire just to wire the spark plus, fired by the magneto. It took 44 feet more to wire the distributor side of the ignition. The oil pan held five gallons. On the road, gas mileage averaged two to three miles per gallon. The small V-12 paled by comparison. It weighed 1,640 pounds, but averaged better than twice the miles per gallon.

The photos, at right, show the Pierce-Arrow style engines that Seagrave used. In 1936, Seagrave entered into a contract with Pierce-Arrow, a producer of fine automobiles, to build a truck version of its passenger car, 8 and V-12 engines. Changes included a different block configuration to adapt to Seagrave's frame and truck style transmission. There were also changes to the cam and manifolds.

Plagued by faltering sales (only 17 big Pierce-Arrows, in 1938) and a mini-depression, Pierce-Arrow declared bankruptcy. Seagrave moved manufacturing operations of their engines to Columbus, Ohio.

The eight-cylinder was Seagrave's base engine, through 1951. The engine's displacement was 385 cubic inches. Its bore and stroke was 3 1/2 x 5 inches, producing 150 horsepower.

The V-12 was called an E-66 by Seagrave (E=engine, 66=two sixes or twelve). This Pierce-Arrow style engine lasted well into the 1960's. The motor was originally produced in two sizes, a 3 1/4 x 4 inch version of 160 hp, and a slightly larger version. The 3 1/2 x 4 inch configuration of 462 cu. in. originally produced only 185 hp. Changes in carburetion and valve train, by Seagrave, quickly increased horsepower to 202.

The B model pump made the transition from the full grille to the anniversary series without many changes. There were two size castings for the main body. One casting was used for the 750 and 1,000 gallon versions, and a separate, but similar, casting for the 1,250 gallon pump. The difference in the 750 and 1,000 gallon pump was 3/16 of an inch width, in the impellers, and more horsepower.

Seagrave constantly tried to improve the pump, even to the point of trying a competitor's impellers. The two stage Seagrave pump was the best, it had no equal.

After a production run of fifteen years, the full grille era came to a close with the last U. S. delivery, a 1,000 gpm canopy cab for Lansing, Michigan. That delivery was on May 15, 1951. The change over at Columbus, Ohio, was abrupt. This was the first Seagrave model change that didn't take months or years to complete.

During the styling change, the driveline and cowl remained the same, but the sheet metal, trim and even the dash changed. The controls became easier to read and more driver oriented.

One option that ceased, with the full grille style Seagrave, was the base eight-cylinder engine. The last U. S. delivery went to Reading, Massachusetts, in 1951. Adoption of the Pierce-Arrow style engines generated many rumors. To quell these rumors, Pierce-Arrow President, A. J. Chantor, on October 16, 1936, issued a statement, "The drawings, patterns and tools which are used in the construction of these engines are property of The Seagrave Corporation and not used in the production of our stock engines." Although some parts of the Seagrave and Pierce-Arrow engines would interchange, they were more like cousins than twins.

The Anniversary Series

1951 - 1970

The pilot car aptly carried the number 1 on its doors

This model had been in the planning stages for seven years. Under this sleek modern skin, many changes in driveline components would take place. Commercial gas or diesel engines, air brakes and alternators would all be options by the end of the Anniversary Series.

The Seagrave Anniversary series, or more correctly the 70th Anniversary series, was introduced in 1951, the 70th anniversary of the founding of the company. The initial design work was actually started, in 1944. Modifications to the basic driveline were at a standstill during the war years. But, the designers sketch pads were busy. The Anniversary Series was being developed along with the wide cab design for the post-war full grille model. Modifications and improvements culminated, in 1951. The first two completed Anniversary Series units were delivered to the Columbus Fire Department for validation, in January, 1951.

Carried over from the previous design were the basic engine, pump and cowl sections of the cab. The criteria of a sound layout was also retained; good weight distribution, low level suction hose attachment, low center of gravity and ease of operation.

The rumor mill carried so many false reports of the new apparatus, from Seagrave, that the factory had to issue a letter in December, 1950, setting the record straight. "In spite of the fact that our competitor apparently thinks he has the best thing on wheels, we will let him have it on an exclusive basis, as far as we are concerned, because we find that many cities will not accept his driver-ahead-of-engine type." And so it stood, for almost another decade.

The sales staff requested to stay with a sure thing. Cab forward was still in its infancy and could be a hard sell. The trend to cab forward grew, however, and the sales staff would reverse itself a few years later. The cab forward became an optional Seagrave design, in 1959. By 1962, the cab forward pumper design outsold the conventional design 84 to 43. In the last year of production, 1970, total Anniversary Series production could be counted on one hand.

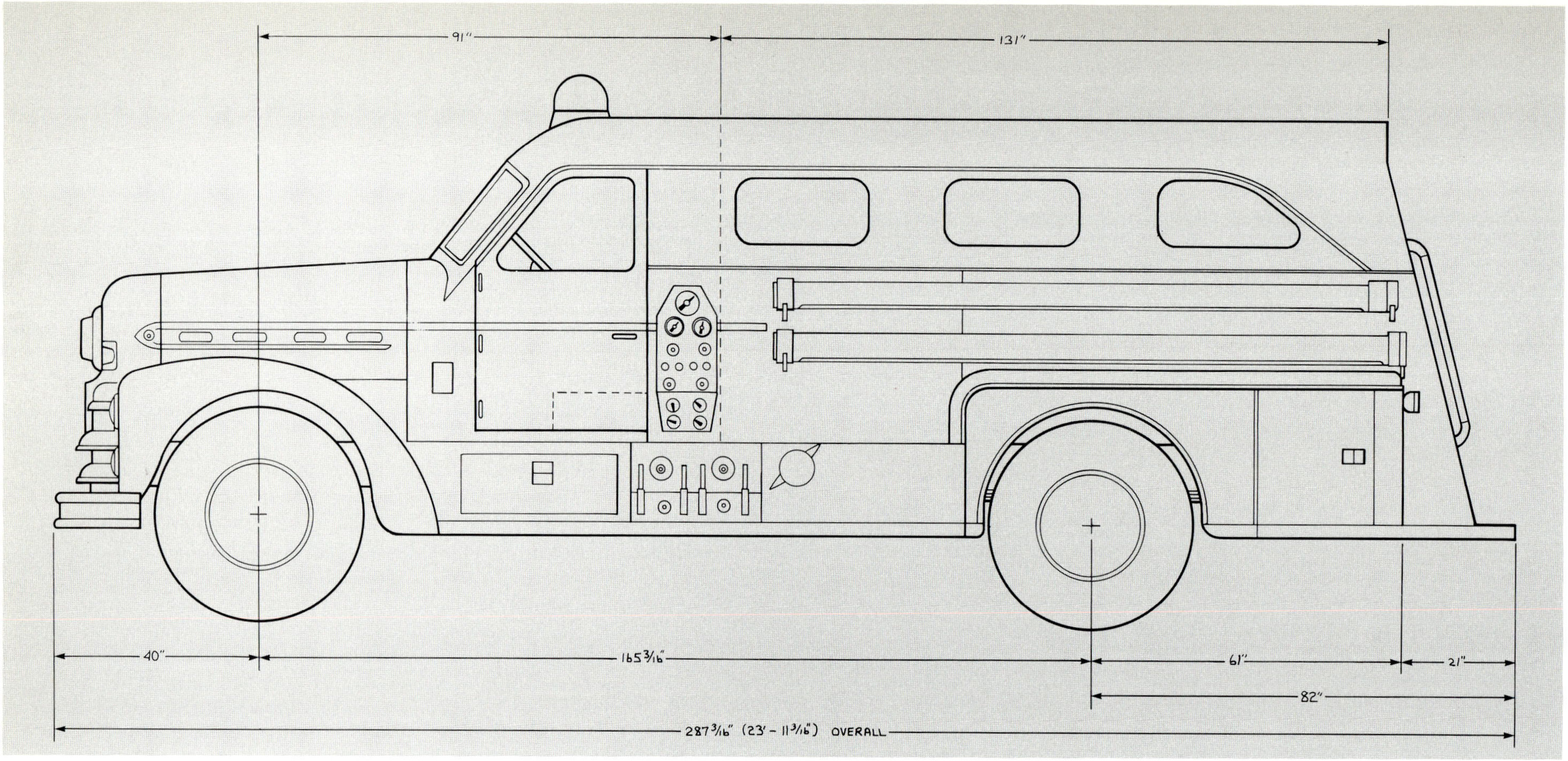

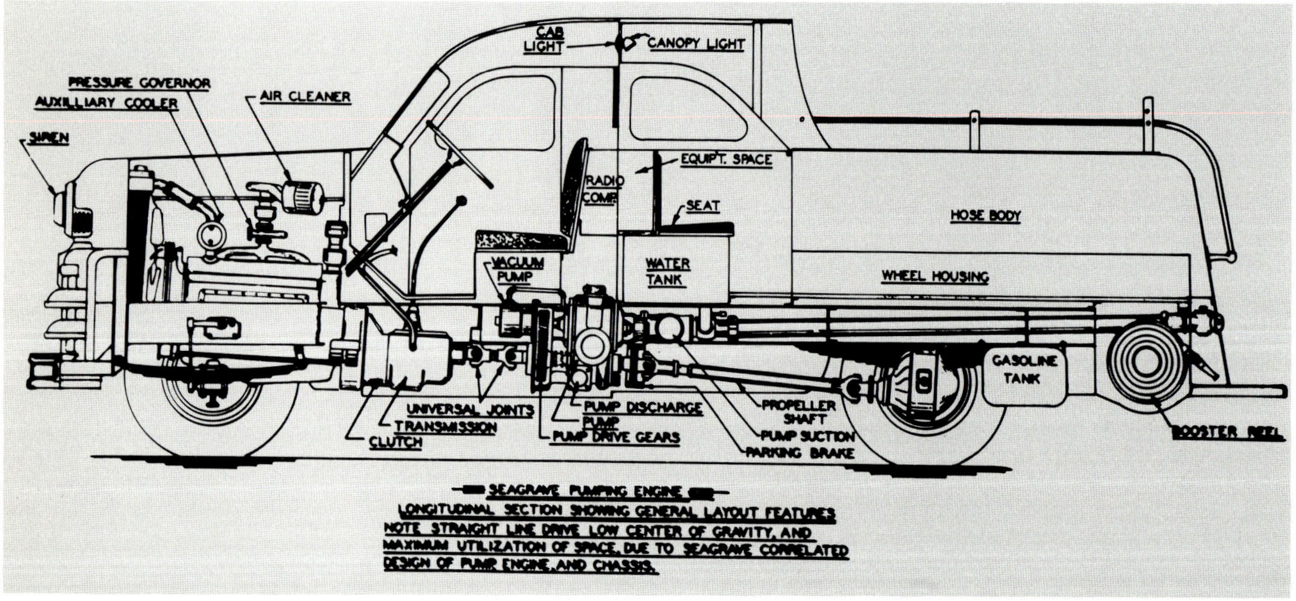

With the pump nestled low in the frame and the even weight or load distribution, the anniversary series was a dream to drive. The pump was a carry over from the previous series and the cab structure was a close cousin to the full grille series. With so many desirable carry over components, and a stylish new look, sales soared. Actual deliveries began in mid 1951.

Some communities that ordered the earlier style apparatus received the new anniversary style. Detroit, Michigan, ordered five pumping engines on October 27, 1950. When the order arrived in mid 1951, four were the full grille style pumpers, the fifth was a modern design anniversary series.

These sketches, detail the dimensions and features for the new style deliveries.

These little mock up models were completed in 1943, according to notes in the photo log. They sported different grille styles that were not adopted for the final product. Surprisingly, one featured the square fender design that would not be adopted for thirteen more years.

There were improvements in more than style. Along with the style came:

- A separate adjustable driver's seat
- Easy service roll out battery tray
- Improved flip-over ladder locks
- Spring loaded suction hose clamps
- Easy-to-read pump panel
- Quarter turn, modified ball discharges

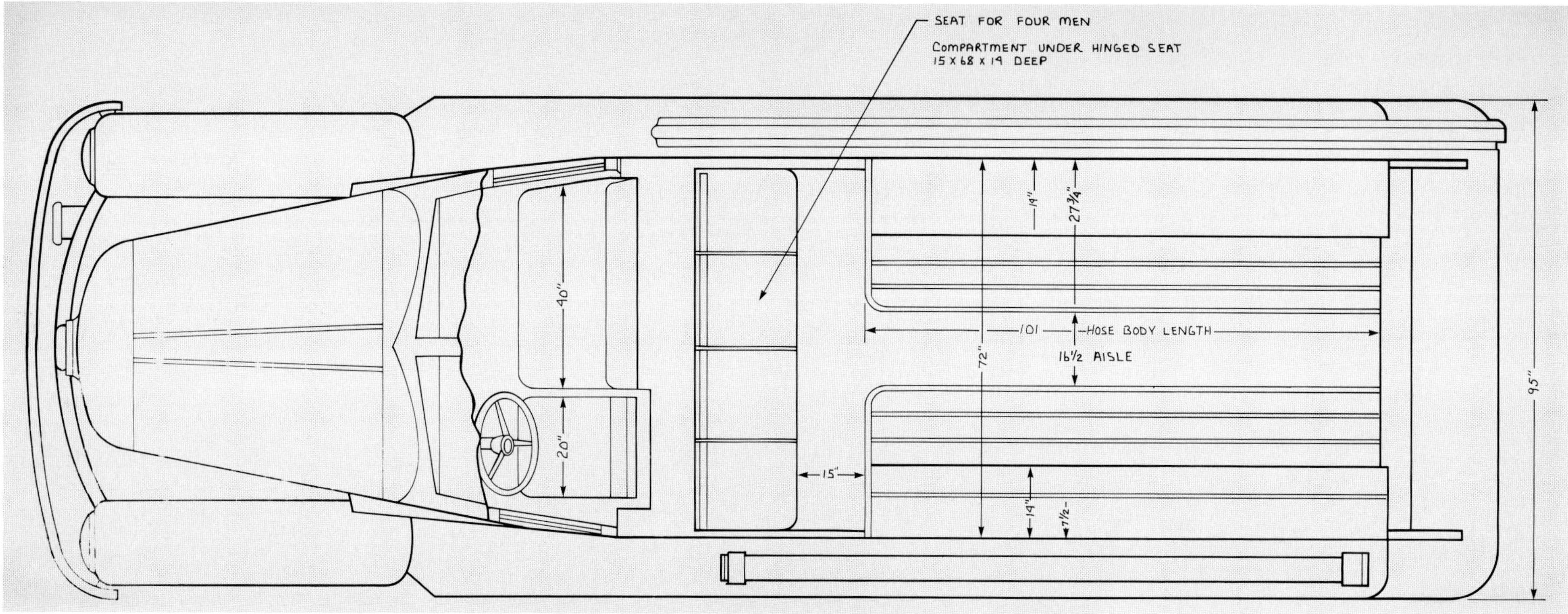

The sketch, above, shows the separate driver's seat, new for this series. In previous models, the driver's seat was fixed. It was not uncommon to see a short driver, of earlier equipment, seated with a large pillow behind him so he could reach the pedals. The area behind the driver and officer's seat could be utilized for crew, storage or water tank. On some deliveries the water tank became so large, the crew had to stand in the middle aisle (note the photos on pages 270 and 271).

The photo, at left, shows the first ladder and pumper. This publicity photo was shot at the new station, Engine 21, at Columbus, Ohio. The ladder was serial #F2140, the engine serial #F2135. This would be the new look at fire stations across the country for the next 19 years.

The set of photos, above, start out with the first anniversary series engine delivered outside Columbus, Ohio. Serial #F4750 went to Garfield, New Jersey. The balance of the page demonstrates the variety of custom body styles that filled the marketplace in the early 1950's. Custom, custom yes; at Seagrave, almost every apparatus looked different. Each delivery was to fit a different purse and purpose. In total over 1,500 Anniversary Series apparatus were produced.

Subtle changes can be noted throughout the 19 year run of the anniversary series. The Pittsburg, California, engine was ordered January 5, 1951, and featured the round fender design that was typical, until late 1957.

Kalamazoo, Michigan, ordered this squared fender, single headlight pumper on January 30, 1957. Serial #5585 was typical of 1957 through mid year 1958 deliveries.

Seagrave fenders were made in the plant. The square cornered fender used on the 1957 and later editions were made in sections. The actual corner was die formed and then welded into the hand fabricated fender.

Open rigs always looked sleek, but cab stability was always a problem. Without upper support from the roof section, the cab sides flexed with each twist of the frame. The result of some twisting was the unexpected opening of the half doors.

A federal law dictated the use of the dual/quad headlight system (separate high and low beam bulbs), for 1958. Auto makers introduced the headlight change at the start of new model production for the model year. At Seagrave, it was a mid calendar year change. Brook Park, Ohio's, big V-12 powered, 1,000 gpm pumper was ordered June 17, 1958, and featured the new headlight system.

In 1963, subtle changes started to show in the body design. The 1969 delivery to Glenview, Illinois, one of the last canopy cabs, shows a higher windshield opening and a taller opening in the front fenders. The warn-ray cowl lights by Guide were also replaced with Unity warning lights. These changes coincided with the move to Clintonville, Wisconsin, in the 1963-1965 time period.

At Clintonville, there was not a brass foundry on site. Because of this, some of the extra brass castings were eliminated, such as, the cast nameplate in the center of the grille.

Shown, on this page, are the most common cab styles during the early 1950's. The open cab accounted for thirty percent of the market. The canopy cab, in various configurations, had fifty percent of the market. The remaining twenty percent was split between the closed cab versions.

The 1956 Pensacola, Florida, delivery was serial #J6975; Tampa, Florida's, 1957 delivery was serial #K905; and Ravenna, Ohio's, 1957 delivery was serial #K6300.

The square rear fender provided space for more compartments. The size and shape of this storage area changed with the wheel base of the vehicle. The old rounded rear fender remained an option.

Seagrave anniversary pumpers came in five series: 400, 500, 531, 550, 800, and 900. The series was based on engine size:

400 =	202 hp	V-12
500 =	215 hp	V-12
531 =	251 hp	V-12
550 =	250 hp	Waukesha
800 =	300 hp	Waukesha
900 =	300 hp	V-12

The model was dependent on cab design. The base design was a three man open cab. But, the options and prices in 1960 were:

3-man closed	Add $650
7-man open	Add $275
7-man canopy	Add $950

Other sample prices included:

531 series	750 gpm pump	$21,750
531 series	1,000 gpm pump	$22,250
900 series	1,000 gpm pump	$24,900
900 series	1,250 gpm pump	$25,400

Major options were:

200 gpm booster pump	$ 850
High pressure pump	$1,350

The list of required features would be spelled out and sent to engineering to make sure it would be possible to build to these specifications. (This would eliminate an unreliable match: example; 1,250 gpm pump with a small engine.) The specification engineer also planned for the incorporation of special features and trim items. Most departments requested some changes. Detroit, Michigan, for example, had no water or booster lines on any of their pumping engines, during the anniversary series.

Getting to the fire was often only half the problem, finding an available fire hydrant was sometimes the other half. The resourceful Columbus, Ohio, fireman promptly solved this urgent dilemma.

The official name for this style pumper was limousine cab. Seagrave made ninety one, fully enclosed, pumping engines, through 1965. Sixty seven units went to Detroit, Michigan, and twenty-four to other communities. Favored in cold climates, sales stretched from Camden, Maine, to Danville, Illinois. The 1952 delivery to Atlantic Highlands, New Jersey, is shown left.

The extra charge for these fully enclosed bodies over a regular canopy cab body was $1,500. The whole crew could be afforded protection from the weather, and the almost roomy interior was a welcome break on one of those long nights guarding downed wires. Today's enclosed ten man cabs harken back to these early sedan days.

Bickle Seagrave, the Canadian partner, completed Seagrave chassis for the Canadian market through their plant in Woodstock, Ontario, beginning in 1936. They marketed a product that, at times, varied quite a bit from the stock Columbus deliveries. In 1952, Bickle Seagrave moved into new and larger facilities on Dundas Street, Woodstock. The company prospered briefly, then floundered, caused by an economic slow-down in 1955 and a long strike at General Motors which supplied chassis for the commercial pumpers. Bickle, or Bickle Seagrave, folded exactly fifty years after its inception.

The pumper for Scarborough, Ontario, near Toronto, was a 1952 delivery. It was serial #F6694. The pumper, shown right, was the last Bickle Seagrave, Canadian serial #J50, a 1955 delivery, a combination foam and water pumper.

The later, 1954 and 1955, Bickle Seagrave deliveries wore the unique center grille medallion. All Bickle deliveries were painted and striped at the Woodstock plant. Most received the attractive accent to the fender openings and an additional body stripe.

The last Bickle Seagrave was this 1955 delivery, a foam pumper for the Imperial Oil Company refinery at Sarnia, Ontario. There were 206 custom Bickle Seagrave deliveries, through the years 1937 to 1956.

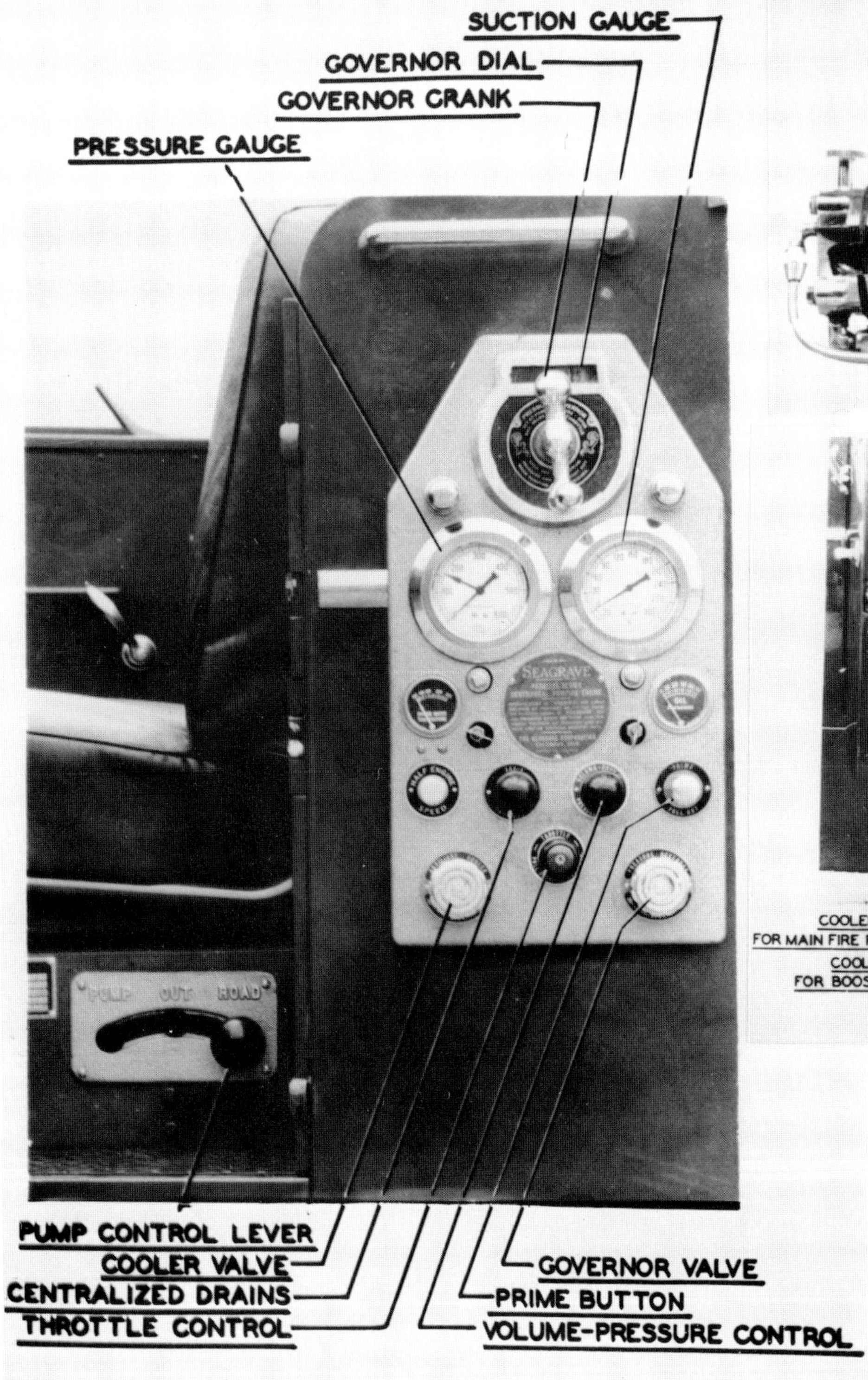

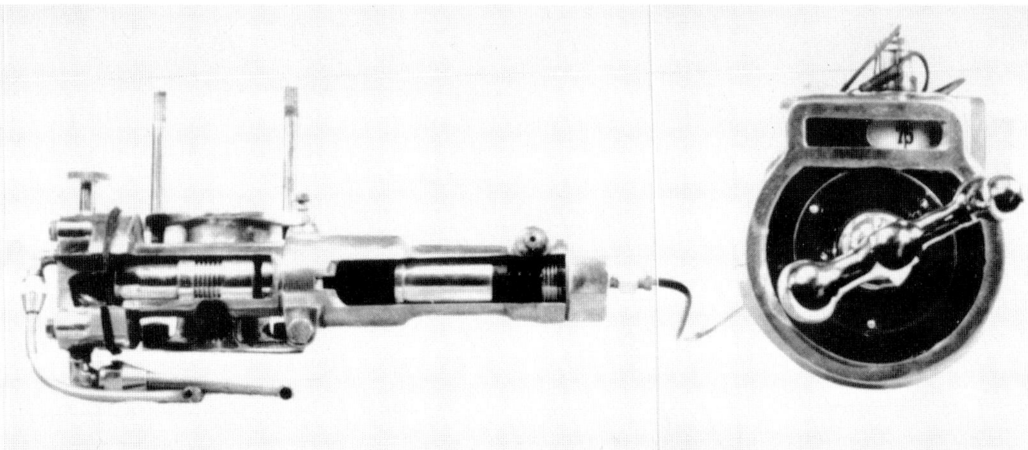

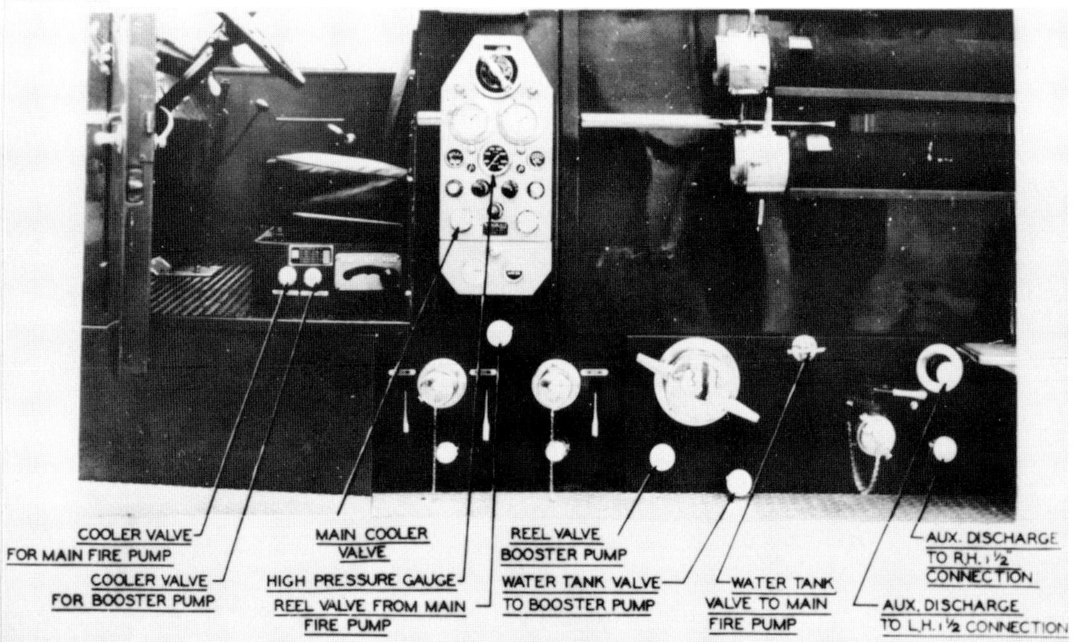

TYPICAL HIGH PRESSURE PUMP CONTROLS

Pump panel configurations changed very little from the panels of previous Seagrave deliveries. Shown here are the panels for the standard pumper, left, and the high-pressure, optional, pump panel above.

The top photo shows the components of the automatic pressure relief system. The system was in two parts, the sender mounted behind the crank handle and the receiver mounted under the carburetor. The sender worked like a master cylinder in a brake system. The more turns of the handle, the more hydraulic pressure was sent to the receiver. It then opened up fuel mixture flow to the engine. The opposite end of the receiver contained a tube, back to the pump. This allowed pump pressure to react to the preset sending unit pressure.

Booster pumps, rated at 200 gpm, and high pressure pumps were both driven by power take offs, from the transmission, as shown right. Hydraulic pumps to operate the aerial hoists were also transmission mounted. Most heavy truck transmissions have facilities for two power take offs.

The high pressure pump was a four stage centrifugal pump. It was constructed of bronze and stainless steel. The inlet to the pump was 1 1/2 inches in diameter. This volume of water allowed the pump to deliver 100 gpm at 200 psi, and up to 40 gpm at 1,000 psi.

High pressure fog, to fight fires, was popular from the late 1940's through the 1960's. Its advantage was that a small amount of water could be vaporized. The water vapor would deprive the fire of oxygen and extinguish the flame. The problem with the system lies in the fact that the combustible material could reignite if oxygen returned before the ignition temperature had dropped sufficiently.

In 1956, the small block V-12 engine received some updating. This new 531 engine is distinguishable from the old 530 by the addition of the Seagrave script to the heads.

The old 530 engine had Seagrave in block letters and allowed a compression ratio of 5.5:1. The new script heads allowed a ratio of 7.9:1. The increased engine efficiency was then matched by change in coolant flow. The flow now was from the top of each head to the radiator tank. The heat exchanger for the pumping system was moved to the lower radiator hose. These, and other changes, raised horsepower from 215 to 251. Similar changes in the big V-12 brought its horsepower up from 268 to well over 300. The 906 and 531 engines would continue to be offered, until 1970.

The basic size of a power plant and its peculiararities (engine mounts, water outlets, etc.) call for varied chassis and sheet metal modifications. The three apparatus, shown here, have three different power plants.

The squad truck for Ridgefield, New Jersey, had a small V-12. It is on a 165 inch wheelbase and has the small hood. Note the hood's length in relation to the fender flair.

Beckley, West Virginia, had a big V-12 on a 174 inch wheelbase. The additional nine inches is all in the hood length.

The Hall Scott six-cylinder powered apparatus for San Diego, California, featured a long hood with special louvers.

The Hall Scott Engine Works was originally of Berkeley, California. They offered engines from 707 to 1,090 cu. in. displacement. The larger sizes over five feet long, from fan to transmission mount. To satisfy its west coast customers, who liked these big six-cylinder engines, Seagrave would modify the basic frame and sheet metal.

The factory offered six-cylinder engines, by Waukesha Motor Company, Waukesha, Wisconsin, as alternate power plants. The Seagrave 550 series received a 250 hp (one less than the small Seagrave V-12) six-cylinder. The 554 cu. in. engine was Waukesha model 140GZB. The 800 series Seagrave used a 300 hp (the large Seagrave V-12 was listed simply as 300+ hp) six-cylinder Waukesha. The larger engine's displacement was 817 cu. in. and its model designation was 145GZB. These engines fit under the standard Seagrave hoods.

Every engine had its downfall. The small Seagrave V-12 labored under the weight of a ladder truck. The large V-12 got only three to five miles per gallon. The Hall Scott engines exhibited excessive timing chain noise. The Waukesha engines suffered casting fatigue. All of the gas engines ran hotter than the diesel engines that would eventually replace them.

The delivery to San Diego, California, was unique for its lack of doors and the box mounted on the rear handrails, that held a bundle of attack hose.

VALLEY STREAM
NEW YORK
4
Seagrave
P4070
Seagrave

ENGINE CO.
CLEVELAND
FIRE DEPT.

Some departments order plain, some fancy. Valley Stream, New York, ordered the chrome dome grille. Then they opted for Mars 8-8-8 lights and subway straps for the crew on the rear tail board.

Madison, Wisconsin, purchased the entire "chrome dome" option; radiator, shell, hub caps, windshield frame, top of door glass, gage panel, switch panel, hood hinge, and grab rail. Cost for the package was $250.

Cleveland, Ohio, on the other hand, bought very little trim, but spent money on the special side entrance crew area. The 16 1/2 inch wide entrance led to a full width, rear facing bench seat accessible from either side.

Columbus, Ohio, was home for this stock model Seagrave. Officials from Seagrave would often invite visitors to local fire stations to view sample apparatus.

Special body builders purchased kits, or partially completed chassis, from Seagrave. Local politics often demanded a special body or a special body builder.

Special order, cab chassis, special engine - Seagrave design staff handled it all. The kit truck, serial #850, was off to the C. N. Johnston Body Company to receive a special body. The completed rig would be a mate to the engine shown, at the lower right. The enclosed squad was finished by the Gerstenslager Company for the Syracuse Fire Department.

The West Milwaukee Fire Department ordered a 1,250 gpm canopy cab with almost every option in the book. The big V-12 powered pumper featured a deck gun, skylight windows, road sanders, high pressure pump, and a complete set of gauges, one for each outlet. Each piece was fitted with the same care as in lettering the hood. The skylight windows on this 1963 delivery became a desirable option . . . so much so, the design staff increased the height of the windshield the following year.

Closely tracking customer's preferences, and the competing brands, helped Seagrave stay in the main stream. The company also hired professors, from the Engineering Department of Ohio State University, during the summer. The influx of ideas from the college professors helped shape new designs.

The factory photos, shown in this book, were taken to document what the truck looked like when it left the factory for insurance purposes, and for ease in identifying repair parts. Without pictures or part numbers, identification of even the simplest part number would be tricky.

The factory has some, but not all, records that go back to 1913. Some early records were transferred to microfilm, and they have not stood the test of time well. The key to ordering parts or obtaining information is the original serial number or the year and original destination.

The apparatus shown here were for Jersey City, New Jersey, in 1958, and Milton, Massachusetts, in 1960. Both rigs featured grabrails, over the hose box, to aid in reloading fire hose.

From 1958 to 1964, Seagrave offered this short coupled type conventional pumper. The Shorty was 21 feet 8 inches long, about 3 feet shorter in overall length from a cab forward pumping engine. Overall width was 91 3/4 inches, or about 3 inches narrower than the standard pumper. The narrow width was accomplished by keeping the gross vehicle weight down and using a lighter and narrower rear axle.

The Shorty could rival most cab forward apparatus, in bumper to wall turning radius, at 27 feet. The cab forward apparatus always claims a tight turning radius, but its long front overhang detracts from its actual turning radius. The bumper to wall turning radius is more representative to actual on road conditions.

Shorty pumping engines came with either 750 gpm pumps or booster pumps. The idea of a short standard pumper was short lived. Most people purchased a conventional style pumper for the handling quality of the vehicle and not the maneuverability in tight places.

The cab mounted nozzle on this truck sprayed foam. The Rockwood foam system, on a Seagrave, became known as the Vigilante series. The rig carried 600 gallons of foam and 400 gallons of water in its tanks.

The Rockwood automatic pressure system proportioned the liquid foam by means of a 75 gpm positive displacement pump. The foam was used mainly in fighting oil fires. The big users were airports and oil handling facilities.

Airports have special needs. Huge structures, crowded facilities, fuel storage and electronic equipment galore. The huge crash truck always gets noticed, but the support rigs are just as important. Seagrave supplies the support equipment at many airports as demonstrated by the CO_2 fog rig for the Orlando, Florida, airport.

Many cities reordered year after year. The cities depicted here were New Haven, Connecticut; Montreal,Quebec; and Detroit, Michigan.

The federal civil defense program channeled money to state governments, throughout the 1950's, to purchase supplies that would be necessary in sustaining a community at a time of natural disaster or enemy attack. Many states purchased fire apparatus under this program.

The commercial pumper business was augmented by many purchases from industry. Commercial chassis were purchased from Columbus area dealers and then outfitted to customer demands. The big GMC chassis could handle a volume pump of 500 gpm with ease. The base GMC engine, in trucks this size, was six-cylinder of 270 cu. in. displacement and 140 horsepower.

The engine in the Chevrolet was a 108 horsepower, 235 cu. in., six-cylinder engine.

The base engine in a Seagrave of this vintage would have a horsepower of 202.

The Ford truck was the choice of most apparatus builders. For Seagrave, it made no difference. Seagrave pump impellers stayed the same size for each unit. Only a gear change, in the pump transmission, was necessary to match up to the power plants torque curve. Other manufacturers had to change pump impeller size to line up with different torque curves, a more expensive and limiting process.

The Ford 6 reached maximum torque at around 1,000 rpm, the V-8 at around 1,900 rpm. It is the combination of horsepower and torque that gives an engine its pumping ability.

UNADILLA
P4060

MANASSAS
VOLUNTEER
FIRE CO.
ENGINE
No 13
M4600
CORP

KEYSTONE HEIGHTS
2
FIRE DEPT.
CORP.

TAFT FIRE DEPT.
NO. 4
CORP

CARY
FIRE PROT.
DIST.
Seagrave
N8270
CORP.

Commercial chassis made inroads into the fire apparatus business, during the 1950's and 1960's. International bragged, in 1957, that its heavy truck had outsold all other heavy trucks, for over 25 years. It offered: standard duty, black diamond engines in 264, 282 and 308 cu. in. displacement; medium duty, red diamond sizes of 372, 406, 450, and 501 cu. in. displacement; and heavy duty V-8 sizes of 401, 461 and 549 cu. in. displacement.

Chevrolet offered seven engine sizes from 216 through 348 cu. in. GMC offered engines from 228 to 702 cu. in. The latter was actually two 351 V-6 engines, coupled together to produce a V-12. Their V-12 engine allowed GM to be competitive when the bid specifications called for V-12 engine power.

Ford introduced the C series tilt cab, in 1957. Within two years, it dominated the commercial apparatus market. The smallest series C550 Ford tilt cab was powered by a 292 cu. in. engine. Engine sizes increased to the 539 cu. in. engine of 277 horsepower in the C850 chassis.

Seagrave created the K series pump to mount on top of the frame of a commercial pumper. A customer could now enjoy a bronze Seagrave pump on a commercial chassis.

The all bronze, all Seagrave pump, shown on this page, is a Seagrave model B. This pump was mounted low in the frame, and on some models the inlet tube actually came right through a hole in the frame rail. The model B pump was standard on all custom apparatus. It was available in capacities of 500 to 1,250 gpm.

The apparatus drive shaft is an integral part of the pump. The B style pump is bolted together, from three main sections, horizontally. For major repairs, the pump had to be removed, as a unit, from below the vehicle. The K style pump had a seam in its case so the bottom half of the pump could be removed while still mounted in the vehicle.

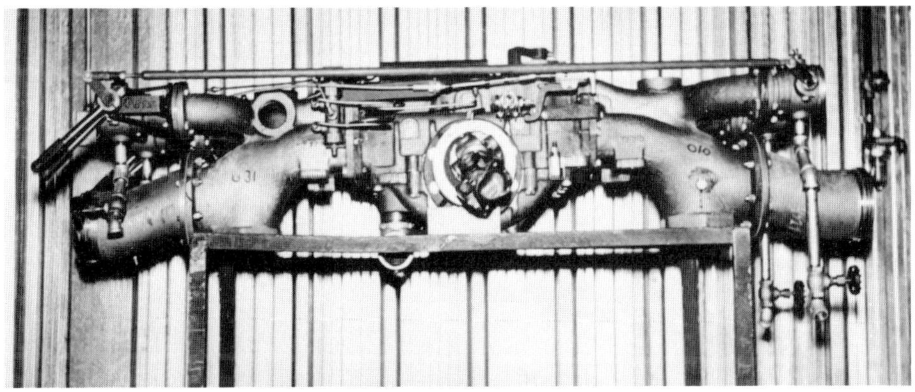

Seagrave's K pump is an all bronze pump designed for commercial or above the frame applications. The drive shaft, of the vehicle, goes under the pump to a separate gear box. The K model pump had dual impellers of 9 1/4 inch diameter. Pump volume was increased by providing a greater water inlet and more horsepower to drive the pump. In the photo above, the dual distributors of the V-12 engine can be seen in the background.

The Anniversary Series ladder truck was more handsome than this artist's rendering. The 1951 Columbus, Ohio, delivery used for this art is still on the Columbus department roster as training truck 'A'. The last Anniversary Series ladder delivery went to Jackson, Michigan, in 1970.

The majority of the cosmetic changes in this series can be best shown in rear view of the Jackson truck. The most noticeable changes resulted in squaring off the fenders and rear cabinetry.

Many communities prefer to combine their pumper and ladder requirements into one apparatus. Louisville, Kentucky, ordered their Quad combination pump and service ladder, in September, 1950. Power for the 750 gpm pump was supplied by the 202 hp V-12 engine, commonly referred to as the "66 Engine". Cost for a quad like this was $19,323.

The Hamtramck, Michigan, Fire Department ordered this tractor-drawn aerial pumper. The tractor pumper, aerial ladder, was not new to Seagrave. The Oklahoma City, Oklahoma, Fire Department ordered two similar rigs, in 1928.

The Hamtramck's apparatus featured: front and side suction intake for its 1,000 gpm pump, the big V-12 engine, a 300 gallon water tank, booster equipment, and a 100 foot aerial ladder. The apparatus carried no 2 1/2 inch hose. While this combination worked good in theory, it was difficult to spot at the fire and close to a hydrant. This apparatus served from 1956 to 1985, but no similar rigs were ordered.

The debate on open vs closed cabs raged on, during the 1950's and 1960's. Probably a greater case can be made for open cab ladder trucks than pumpers. The positioning of the ladder is of prime importance. Saving a life during the early minutes of the fire may depend on proper spotting of the apparatus the first time. Note a portion of the turntable's base is folded up to obtain easy access to the crew seat. The controls for the Seagrave aerial ladder were on the driver's side. The competition had their controls on the officer's side.

Quad ladder trucks, like the delivery for San Diego, California, were popular in one and two story neighborhoods. By 1960, the service ladder truck accounted for less than ten percent of the ladder truck deliveries. Raising a hydraulic aerial was as quick as positioning and raising the hand ladder. The midship mounted aerial ladder truck took over the service ladder market.

The hydraulic fluid is carried in a 30 gallon tank nestled deep between the frame rails. The hydraulic pump is driven by power take off from the road transmission. Very little fluid is used to rotate the cable drum and that fluid is immediately returned to the tank.

The fluid, to raise the ladder, as much as 16 gallons of it, is held in the two lifting cylinders until the ladder is returned to the bed. Safety valves in the base of the lifting cylinders prevent rapid lowering of the ladder in the event of a hydraulic hose or tube rupture.

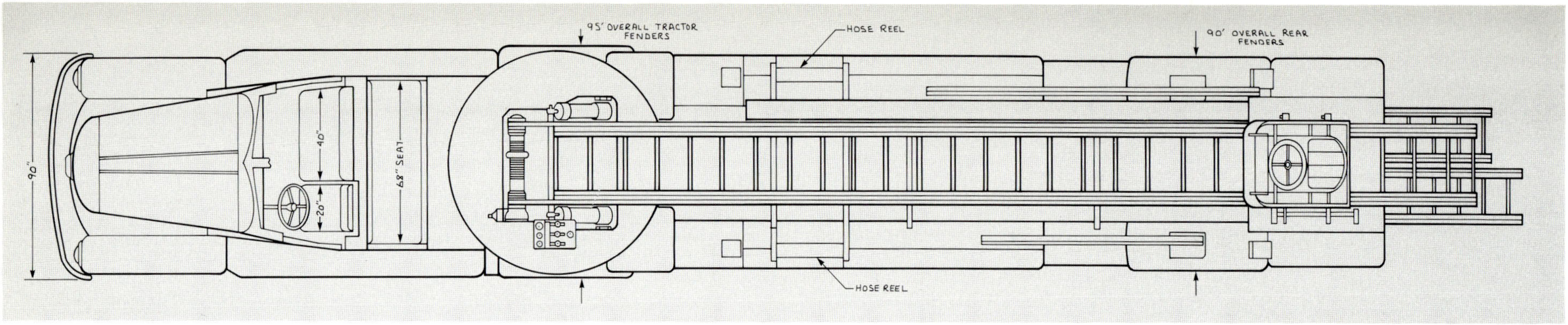

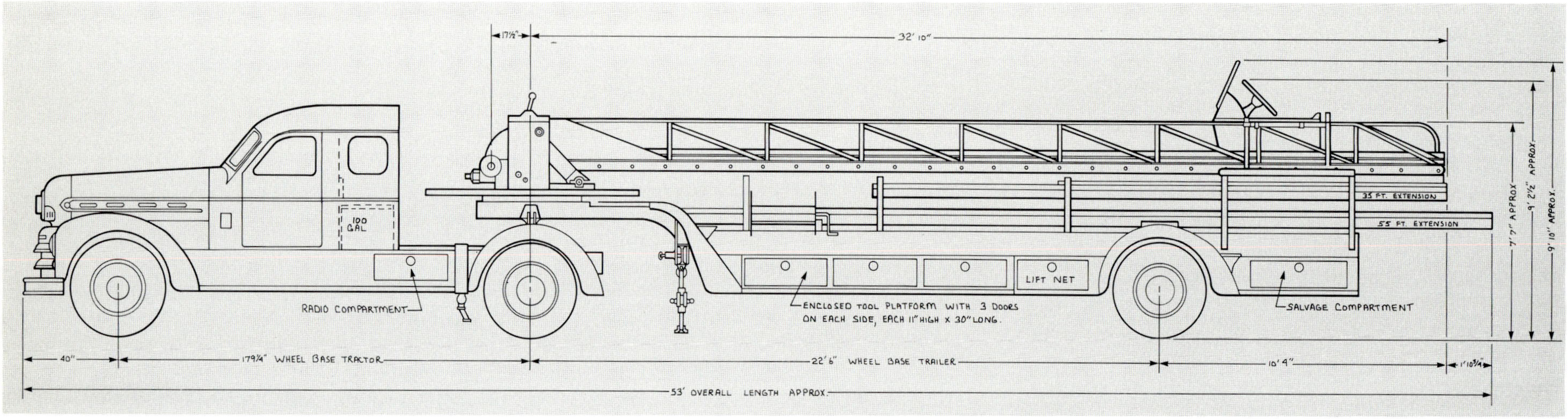

The dimensions given, above, are based on a big V-12 delivery to Detroit, Michigan, in 1955. Overall length would depend on ladder selection, and height would depend on load and tire size. A rig like this would go over 30,000 pounds without add on equipment.

Note the booster reels on the trailer in the drawing above. The addition of two booster reels, pump and necessary plumbing would raise the total weight of a ladder truck several hundred pounds.

The view of Boston's aerials, at a fire, show the difference in fender width and headlight location, between the 1957 and 1958 deliveries. Close examination of Ladder 15 reveals the road sanders and tire chains often necessary during harsh winters.

Las Vegas, Nevada, and Quincy, Massachusetts received new aerial trucks in 1963. Both rigs had their tell tale markings. The Las Vegas truck had dice painted on the splash apron and Quincy trucks had the tractor mounted deluge gun.

The overall length of a seven man cab, 100 foot, three section, tractor trailer ladder truck was 55 feet, yet its turning radius was only 33 feet. That's 10 feet shorter than a straight frame ladder truck. Turning radius may be a vague term, but in the real world, nothing maneuvers, in narrow streets, better than a tractor-drawn rig. Over the years, many cities abandoned this style apparatus to try a trendy item only to return to the "traffic fighter" ladder truck.

This photo, near the right, shows four Boston, Massachusetts aerials. Two of the trucks were 1956 deliveries, the ones with the bells on the grills. The 1957 deliveries had standard bell mounts. The far right photo shows a 1962 Boston Seagrave in action at a fire on Commercial Street.

The Seagrave aerial ladders were designed to carry a load of 300 pounds, on the outer end of the ladder, when fully extended horizontally, cantilever style (see page 224). When placed against a structure at an elevation of 70°, design weights were 150 pounds on every third rung.

The steel used in making the aerial ladders, like this 1965 Lansing, Michigan, apparatus, has a high copper content to improve corrosion resistance. The ladder base stock is of high strength, low alloy steel and is of welded construction. The rig served Lansing until 1989. It is now being restored by the Author.

There has always been a debate on whether a three section or four section ladder is better. Three section ladders are more rigid, because there are less pieces, and the top section is wider and higher. Four section ladders are quicker to erect at the fire. Four sections provide a safety factor of more overlap, section to section. Seagrave 100 foot, three section aerial ladders weigh 2,120 pounds. The base section alone is 975 pounds, the middle section weighs 680 pounds and the fly weighs 465 pounds. Because of the added weight of the extra section, a four section 100 foot, ladder weighs 2,741 pounds.

The choice of three sections or four is up to the buyer or sometimes the tiller driver. The overall length of the three section 100 foot ladder, when nested, is 39 feet, while the four section is 10 feet shorter. The additional gain of 10 feet allows for a fixed tiller seat.

Total time to cycle a four section ladder under full power is 116 seconds. The times include; raising - 14 sec.; extending - 18 sec.; rotating 360° - 32 sec.; retracting - 28 sec. and lowering - 24 sec.

Constant improvement is necessary to build and maintain leadership in the apparatus field. Pictured at the left is the first Seagrave aerial platform. Pictured at the right are a destructive ladder test and the first 100 foot, four section, aerial ladder.

The Seagrave Company's first attempt at an aerial platform was this 1963 delivery to Nashville, Tennessee. The 65 foot boom, was mounted on a special extended chassis so there would be no front overhang. The high center of gravity, created by this design, limited the roadability of the 65 foot Eagle platform. No further articulated booms were mounted on conventional chassis.

Syracuse, New York, received the first four section, 100 foot aerial ladder in September, 1958. The cost of a service aerial ladder truck in 1958 on a small V-12 chassis was:

65 foot	$ 32,750
70 foot	33,650
85 foot	34,750
100 foot	36,250

The open cab ladder truck, shown, is going through reverse stress testing. The Quonset hut in the background held extra wood molds for the foundry. Apparatus waiting for delivery was also stored in this building. It was built in 1948 and was located south of the plant, on the other side of the stone viaduct.

Always innovative, Seagrave designed this light weight aerial ladder, in 1954. It was for commercial vehicle applications or it could be adapted to existing service aerials. The light weight ladder was available in only the 65 foot size. It was lifted by a single hydraulic cylinder, centered between the ladder rails.

The flex of the ladder and its low side rails were not favorable and the design was discontinued, by 1960.

Ladder flex and low sides also doomed the prospects for Magirus ladders in this country. The ladder had six sections and a 146 foot reach. The folks at Green Bay, Wisconsin, took delivery, in 1957, of this one of a kind design.

The service aerial on a commercial chassis was available in 65, 75 and 85 foot versions. The base price for a light weight type 65 foot aerial on a commercial truck, as shown left, was $17,500. A custom type 65 foot aerial ladder on a commercial chassis was $19,500.

The price of the chassis would have to be added to the stated equipment prices. The prices quoted were for chassis modifications, cabinets and Alcolite brand ground ladders. Duo Safety ladders or Seagrave wood ladders were a $250 option.

If customers wanted pump equipment on this commercial chassis, the extra charges, in 1959, were: four stage high pressure pump - $1,750; 200 gpm booster pump - $1,350; and 750 gpm pump $4,500.

The addition of all the items still totaled less than a full custom Seagrave so many communities chose the commercial option. Shown are aerial ladder trucks for Puerto Rico; Bellefonte, Ohio; and a prototype rear mount ladder truck.

The first rear mount Seagrave aerial ladder was fitted to this 1955 Chevrolet chassis. This standard 65 foot aerial ladder was to protect the shipping facilities of the Texas Oil Company at Port Authur, Texas. Rear mount design did not catch on for another ten years.

Some equipment serves many years and survives relatively unscratched. Other rigs seem prone to an early demise.

On July 19, 1966, Engine 8, a 1952 Detroit, Michigan, sedan was destroyed while pumping at a train depot fire. A sudden shift of wind and flame caused the crew to abandon their engine, while still pumping, and seek cover. The vehicle was stripped and the bare frame was used by the Detroit shops as an engine and pump test stand.

This new ladder tractor for Buffalo, New York broke loose from its framing in the box car. The tractor rolled back and forth in the box car for hundreds of miles. Part of the damage to the sheet metal was caused by loose framing timbers.

The photos here graced Seagrave ads. Shown, above, are three deliveries to St. Paul, Minnesota, in 1952. Two Seagrave deliveries are highlighted at a night time fire in Casper, Wyoming, in 1954. The San Francisco, California, ad features a pump and hose car.

San Francisco had two piece engine companies. They consisted of pump and hose car, shown here, and a companion car carrying water and booster pump equipment. Seagrave provided both the pumper and the companion apparatus for San Francisco.

Seagrave did not advertise as much as competing makes, at the time. Seagrave relied instead on repeat sales from long time customers. The big advertising of the day was done by Mack, Pirsh and American LaFrance.

Endicott, New York's, 1967 delivery encompasses some of the quandary of fire equipment of the late 1960's. The engine up front and open cab design was desired by the conservatives. The progressives wanted big water and a lot of storage space. The happy solution was this almost traditional looking pumper with a 1,250 gpm pump.

The end of an era. From 1936 and for 29 years, Seagrave built this safety style sedan cab for nineteen different communities. Detroit, Michigan, received the lions share of sixty-six sedans while Cleveland, Ohio had only eight. The last sedan, a 1965 delivery, is shown at the left. The concept is carried forward today, get the crew off the running board and protect them from the elements.

The last engine ahead of cab delivery went to a long time good customer, New Haven, Connecticut. The last canopy cab Seagrave went to Jackson, Michigan, another loyal Seagrave customer. Both deliveries were made in 1970.

This was not just the end of a series. These deliveries marked a change in design. Since the wagon days, the horse or horsepower had been up front. Most drivers would tell you that the engine ahead of cab rode and drove better than any other style apparatus. But visibility and overall length of an apparatus dictated the change to cab forward. The look of the fire engine has changed.

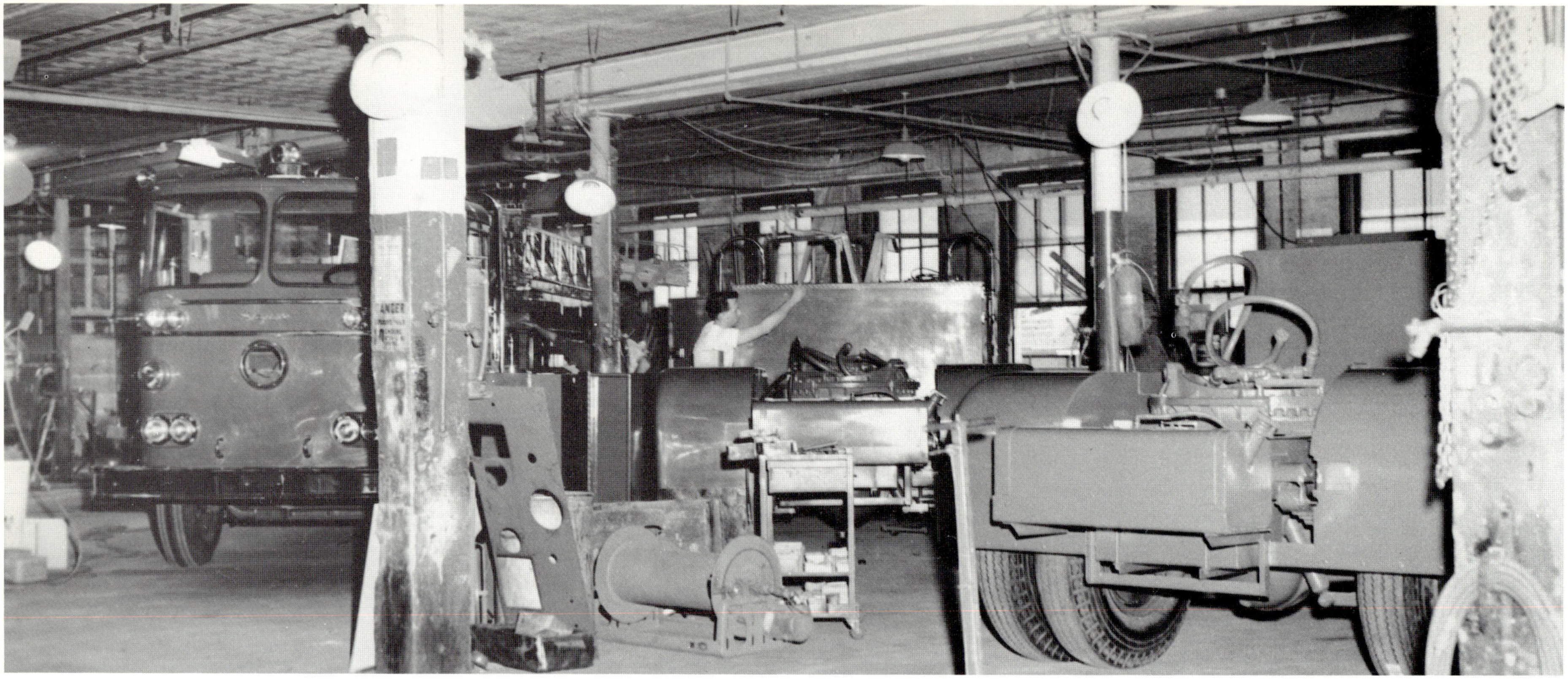

In January of 1961, the backlog of deliveries had reached 71. By January of 1962, the backlog was 81. By the end of 1962, the total was up to 130. The old facility was not keeping up.

The weight and size of equipment being manufactured was many times greater than the first products produced. This management was now faced with the same problem that faced Frederic Seagrave, decades before, orders surpassing facilities. The Seagrave corporate management had a diverse product mix including paint, leather products, housing and two fire equipment companies (Seagrave purchased Maxim Motor Company of Middleboro, Massachusetts, in 1955). The decision was to sell Seagrave and use profits to enhance the rest of the product mix.

Different assembly methods are in use at the Clintonville plant. The apparatus is assembled in a straight line fashion. This 1965 photo shows the main assembly building, unencumbered by posts and a low ceiling.

At Clintonville, the additional Seagrave deliveries more than doubled vehicle production and there were obstacles to overcome. Some of the very first assemblies were sent back to the High Street plant for detailing and road testing. The service section and paint shop were the last to close at Columbus.

The auction notice says it all. The wood working and foundry departments never made the move to Clintonville. Some early records were sold, inadvertently, with the office equipment. The buildings were sold to a local businessman, who subdivided them. The old buildings are still humming today. There are a variety of companies doing basic sheet metal fabrication in the old home of Seagrave.

PUBLIC AUCTION SALE

MACHINE TOOLS • WOODWORKING EQUIPMENT
FACTORY SUPPLIES • OFFICE EQUIPMENT
BRASS FOUNDRY • LAND & BUILDINGS

formerly the property of

SEAGRAVE FIRE APPARATUS, Incorporated

2000 SOUTH HIGH STREET • • COLUMBUS, OHIO

(THIS FINE LINE OF FIRE FIGHTING APPARATUS WILL BE MANUFACTURED IN CLINTONVILLE, WISCONSIN)

SALE DATE: Friday, May 21, 1965
SALE TIME: 9:00 A.M. (C.S.T.)

The management at FWD Seagrave in Clintonville was soon overcome by the flood of back orders that came with the Seagrave acquisition. Management decided to license the use of the Seagrave name for commercial chassis apparatus. Timmons Metal Products Company of Columbus, Ohio, was a maker of truck bodies. Timmons formed Timpco to produce the commercial chassis Seagrave deliveries.

Seagrave stock components were sent to the Columbus plant of Timpco for assembly. From July, 1966 to October, 1969, Timmons provided 140 pieces of equipment. Their largest single order, eight units, went to Austin, Texas, in 1967.

Shown, left, is a pumper for London, Ohio. The ladder truck was for the city of Columbus, Ohio. It featured roll up compartments on one side. The roll up feature was used on Timmons' other product lines, beverage trucks.

The Modern Era

1959 -

The first cab forward delivery went to Jackson, Tennessee, in June of 1959

Seagrave either builds the components for their apparatus or has them built to Seagrave specifications, right down to the bell.

Normal bell metal, for example, is 81% copper and 19% tin. The Seagrave specification called to 78% copper and 22% tin. More tin content makes the bell harder and therefore, ring louder. Lead or zinc which are normal brass additives, would cause a dull tone.

During the mid 1950's, It became apparent that the trend in fire equipment would swing toward the cab forward design. The benefits of better visibility, from the cab, and easier maneuverability in traffic were causing the swing to the cab forward design. Once the design features on the 531 V-12 engine were finished, in 1957, the engineering team began to work on the cab forward design.

Customer surveys and early design validation took almost a year. Fixturing and prototype building brought the introduction date to early 1959. A March, 1959 Fire Engineering advertisement extolled the benefits of the new Seagrave, but the ad contained only an artists rendering of the cab. Actual deliveries did not begin until September, 1959. The development of alternate power plants also took place during the development of the cab forward models. Seagrave followed the industry trends. The goal had never been, be first, just be the best. The diesel engine, the automatic transmission, the trend to ever larger pumps, all kept design staffs busy.

The move to Clintonville, Wisconsin, in the 1963 to 1965 period caused a change in some of the basic design concepts. FWD, a long time maker of quality fire apparatus, in its own right, used more standard components than did Seagrave. For example, they used Waterous pumps and cabs from Truck Cab Manufacturing Company. As the two designs merged over the next decade, the new apparatus would acquire the best of both worlds.

The evolution of the rear mount aerial ladder would call for a new low height cab. Safety regulations would call for a more enclosed cab. And the changes will continue past tomorrow. This seventh chapter in the evolution of the Seagrave product will not be the last. As we go to print the designers are working to surpass the expectations of tomorrow's customers.

The delivery for East Point, Georgia, and all first year deliveries, had a vacuum assisted rod-pump shift lever. The low placement of the pump shift lever was the only concern of the cab forward customers. Within a year, the style of the lever was changed.

Customers enjoyed the visibility provided by the cab forward design, and the ride. With the new model came a Z frame. The top lip of the frame pointed out, the bottom lip pointed in. The Z shape allowed more room in the engine compartment and increased stability.

The back drop for these photos, and many Seagrave delivery photos, was a high canvas drape with a mural painted on it. The drape was supported by wires and was located next to the storage Quonset hut across High Street from the plant.

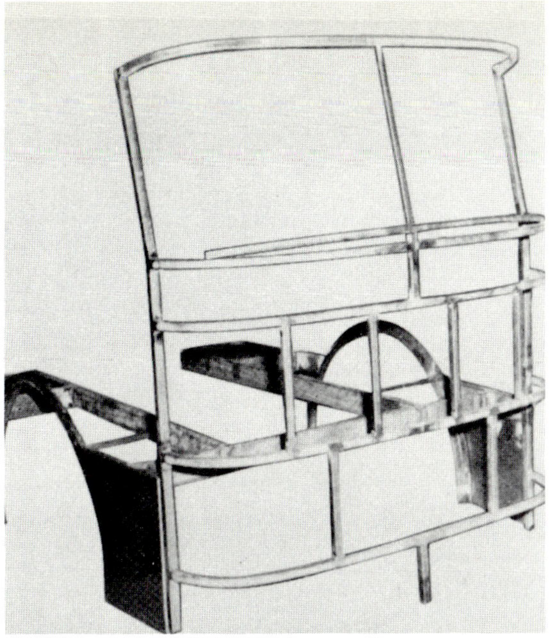

The new cab at Seagrave evolved from the best of the old manufacturing process and what the customers indicated they wanted on the new equipment.

Carried over into the new cab was the structural integrity of the tube steel cab frame. The sheet metal shop foreman created a few mock-ups, then the design and delivery engineers each had their say. Soon, the new cab was born.

Old die hards at the plant quipped that "We'll have to get bus drivers to deliver these damn things." And deliver them they did. In just three short years, by 1962, cab forward deliveries were running 3 to 1 ahead of conventional deliveries. Totals for 1962 were:

Commercial Pumpers	35
Conventional Pumpers	43
Cab Forward Pumpers	84

When you figure that the K Model pump went into the commercial and cab forward models, the K Pump was a 2 to 1 choice over the conventional pump.

The technician in the background is tuning up the engine prior to final assembly. It was a lot easier to get at the engine without the fenders on. After his preliminary tune up, the rest of the body was assembled and a road test performed. Then a final test before delivery. These steps were necessary because many cities had qualifying road tests of their own.

In New York City, part of the delivery test included climbing St. George's Hill. The apparatus had to climb this four block long 12% grade and maintain 30 mph to the top of the hill. In Cincinnati Ohio, the apparatus had to reach the top of White Hill using less than 90% of its rated rpm. The parking brake also had to be able to hold the apparatus from rolling on the hill.

The new dash design had a gage and switch for everything, in sharp contrast to the truck on page 75.

The overall length of a cab forward model was 25 ft. That's about 2 ft. longer than an Anniversary Series delivery. The new model's wheelbase, at 160 inches, was actually a few inches shorter than the previous one. The main difference, then, between the two styles was actually front overhang. Sixty-three inches of cab was ahead of the axle vs forty inches of hood and grille.

By the time Serial #M4000 was delivered to Columbia City, Indiana, in December, 1960, the pump shift lever had changed to this long-lever style.

The customer's desires were met with the same vigor as always. Each delivery could be engineered to the specifications of the community. The early 1960's was the time of the dominate change from open to closed cabs. By the end of the decade, departments were asking for riot proof enclosures.

Base 1960 prices for the cab forward pumpers were: small V-12 - $21,000; big Waukesha 6 - $23,100; big V-12 - $23,900. Minor options were: tachometer on dash - $50; chrome door hinges - $25; fender steps - $20; chrome tow loops - $20; stainless steel running boards and rear step - $400.

The photos, at right, show the testing of a 1,500 gallon pumper for Los Angeles, California ordered in January, 1964. Los Angeles , a long time user of large volume pumpers, requested 1,091 cu. in. Hall Scott power. The plant, in 1964, was backlogged with orders, so these 1,500 gpm rigs were tested at the river.

The cold water of a stream will always produce more pump volume than the warmed water of the typical indoor test tank.

Comparing the photos between the 1964 and the 1970 deliveries reveals many subtle changes. The 1964 models were built to rigid Los Angeles specifications that required a more rearward pump location. Specifications also called for a broom closet, forward of the pump, and access holes in the front of the cab.

By 1970, the pump was moved forward for better weight distribution, the hose box was moved forward, and there was now room for a cabinet ahead of the rear wheel. The old style mechanical siren gave way to electronic, the warning light was now yellow, and the brooms were now stored in the compartment over the rear wheel.

To protect the manned spacecraft center at Houston, Texas, and the launch operation center at Cape Kennedy, Florida, the Seagrave Vigilante was chosen. The 1,250 gpm pumpers also featured a high pressure foam system (see page 276). This special pumper could dispense fog, fog foam or straight foam. The two distinct pumping systems allowed the dispensing of foam and water simultaneously.

Other uses of the Seagrave Vigilante were: Nashville, Tennessee; New Haven, Connecticut: Quincy, Massachusetts; Port Everglades, Florida; and the Dow Chemical Company.

The apparatus on this page represents two firsts; the first Seagrave pumper built at the Clintonville, Wisconsin, facility, and the first factory installed diesel engine in a pumper.

The community fire protection district of St. Louis County, Missouri, chose an 8V-71 Detroit Diesel to power their 1,250 gpm pumper.

The 8V-71 engine was rated at 350 hp, and weighed 2,345 pounds. Its power and weight were a match with the 906 Seagrave that it replaced, so engineering specification changes could be kept to a minimum. The diesel, however, required a 4 1/2 inch longer and 8 inch wider engine compartment. The changes are evident when comparing the two photos on this page. Diesel pumpers on the 164 1/2 inch wheel base are identifiable by the bulge in the lower engine compartment cover.

On the way from the Clintonville plant to Winston-Salem, North Carolina, this pumper stopped at the old Columbus plant for final fit and finish adjustments. It, and the next two dozen or so units off the line, went to Columbus prior to delivery in late 1964.

KING SEAGRAVE Limited

King-Seagrave

With the demise of Bickle Seagrave in February, 1956, the management in Columbus went seeking a new Canadian partner. The logical choice was Vernon Bickle King, a nephew of the original Bickle brothers, who had designed the Bickle/Ahrens-Fox apparatus in the late 1920's. Mr. King had moved from the fire engine business in the 1930's to the delivery truck body and trailer business. By late 1956, King and Seagrave joined forces to produce King-Seagrave apparatus.

The plant produced custom Seagrave apparatus and a complete line of commercial apparatus. They produced 24 deliveries on the venerable old K cab chassis, including the three shown here. The deliveries shown above were for Penticton, British Columbia, in 1960, and Edmonton, Alberta, in 1961. The Calgary Airport, McCall Field, received this 4 x 4 crash truck, in 1963. The King-Seagrave LTD empire collapsed in 1985, when a slow economy led to tight bidding wars on the remaining business. Today, Canadian Seagrave deliveries come straight from the Clintonville factory.

The Cincinnati cab, made by the Truck Cab Manufacturing Company of Cincinnati, Ohio, and the Seagrave cab are often confused. A close comparison of the top two photographs of Cincinnati cabs with the Seagrave cab, at left, will reveal obvious windshield, side window, and door hinge differences. Not so evident was the internal construction and reinforcement of the Seagrave cab.

The cab over chassis from International Harvester was not a familiar item at Seagrave. Many smaller custom builders used the model CO8190 chassis, but builders were limited to the driveline combinations utilized in its frame.

The 1963 deliveries to Buffalo, New York featured two suction intakes. The two intakes were sometimes necessary for 1,500 gallon capacity, as the early K pumps were starving for water.

THE STURDY CONSTRUCTION OF THIS SEAGRAVE CAB SAVED THE LIVES OF FIREMEN

The Mt. Troy Pa., Fire Department's 1000 G.P.M. Seagrave pumper had three men in the front seat, and one in each of the jump seats, plus five men on the tail board. The truck rolled over an embankment. The men on the tail board jumped clear but the men in the seats rolled with the truck. The truck made three complete turns in rolling down the embankment.

Fire Chief Stephen C. Walther writes, "I speak for my Company and particularly for the five men who rode this truck over the hill, when I say, that if it hadn't been for placing quality before price we would not have been so fortunate in having no real serious injuries or even deaths as a result of this accident. I am convinced that the fine engineering and construction that go into your cab was responsible for this. You cannot put a price tag on the safety of your firemen."

The text above appeared in ads for Seagrave apparatus, in 1963 . The built-in roll bar of the Seagrave cab saved these lives. The roll bar, not shown on page 307, is evidenced by the creased roof in the photo at left. No doubt those firemen that rode the apparatus over the hill that day were glad that Seagrave pioneered this safety construction almost 30 years earlier. The pumper was forced onto the soft shoulder of the road by traffic conditions. Getting to the fire is often more dangerous than the fire itself.

The Seagrave six man cab, as the promotions referred to it, was actually a stock FWD tractioneer chassis with Seagrave fire equipment. The idea behind the FWD cab was to get a utility company's entire line crew to the work site at one time. The Seagrave version, called the Optioneer, gave the customer an option: full custom, or custom cab forward with a reduced price. The Optioneer was not well received, owing to the fact that the occupants of the rear seat had to crawl across the front seats to exit. Corry, Pennsylvania's rig was a 1966 delivery.

The cost of producing a small number of apparatus on a custom chassis is not economical when compared to using stock chassis. Communities on a small budget have asked for a custom apparatus with the frills thrifted out, like the Seagrave Optioneer. But, costs of producing a small number of budget apparatus prohibitive.

Clintonville Fire Apparatus Company, C.F.A., built fire apparatus using Seagrave components under license from Seagrave in the 1970's. C.F.A. built its first pumper, for Wichita, Kansas, in November, 1971. The production facility employed fifteen persons, and the product was a commercial chassis with Seagrave equipment.

The C.F.A. factory was a few blocks away from the Seagrave factory. The production at C.F.A. averaged two or three units a month. Seagrave would provide pumps, transfer cases, even complete ladder assemblies. All the cabinetry was done in the C.F.A. shops. The two companies were separate businesses. A few of the later C.F.A. deliveries even included Hale pumps. When Seagrave dropped their own old style pumps, if favorable of Waterous pumps, C.F.A. switched to Hale pumps.

During the economic downturn in 1978, orders for all types of fire equipment dwindled, and marked the end for C.F.A.. Some of the employees moved over to Seagrave and the former factory is now used to produce utility trailers for the U. S. Government.

The history of the Ford tilt cab is a part of the history of Seagrave. With a production run of 33 years, many departments received Seagrave equipment on this chassis. The composite photo, at left, shows the last and first year's deliveries of the Ford truck, commonly called the C-Cab.

Cab forwards were nothing new to Seagrave, they started out with the buckboard style apparatus. And the tilt cab had been produced by both Sterling Motors and White prior to the Ford version. But, the Ford C-Cab was the most well received tilt cab truck ever built. Ford provided the design and tooling for the cab, but it was built by a supplier, the Budd Company.

This cab was available to other truck manufacturers under an agreement from Ford. It had appeared as the Mack N-model, and FWD has used it on a custom chassis. On non-Ford applications, extensive changes had to be made to the floor pan.

The cab underwent minor changes, mostly the trim. The crest disappeared in 1974. The block Ford letters gave way to the familiar Ford oval with the 1985 model year.

Seagrave's first Ford C-Cab was a 65 foot aerial ladder truck for the Fire Chief's Convention in Los Angeles, California, in September, 1958.

The K style pump and Ford C-Cab became an instant success. Noted fire apparatus historian, Walter McCall, calls the Ford C-Cab apparatus a "belly button truck", (everybody has one).

Over time, the cab was modified by Seagrave to become a five man cab, like the 1979 delivery to Corvallis, Oregon, shown here.

When ordering a custom or commercial chassis apparatus, most communities adopted proven designs like the one shown at right. They then made small alterations, if necessary, to meet local demands. Selecting from a host of proven designs is more cost effective than starting with a clean sheet of paper each time. Even if the cabs of these commercial deliveries all looked alike, Seagrave custom built each apparatus to proven design and customer requirements.

Q79336
Seagrave

Seagrave
R79482
Seagrave

Seagrave
Seagrave

The Seagrave Invader series was created to fill the demand for a true custom apparatus at the lowest possible price. This was a true, no frills, fire engine. Plain square sheet metal and a Spartan look to the interior helped keep the cost to a minimum. The Invader series was priced to fill the gap in cost between a custom apparatus and a class 8 commercial delivery. A total of 82 Invader or "I" cabs were built. The U. S. Navy took delivery of 42 units, the remaining 40 went to communities. Beaver Falls, Pennsylvania, received the first civilian delivery, in August of 1983. The Philadelphia Fire Department received eight of the Invader series apparatus.

Commercial chassis builds at Seagrave were produced occasionally to meet specific customer demands or to prove out market trends. The competition in the fire apparatus market requires that builders keep abreast of developments in the entire truck field.

Trends though the mid 1970's and 1980's revealed the mini-pumper was a very saleable item. Small departments ordered them as front line rigs. Large departments also saw the benefit of their small size. A mini-pumper could be dispatched to a fire in an underground parking garage or traverse soft ground at a park.

The mini style pumper was available on Ford, Chevrolet or Dodge, one ton truck chassis. Serial #M77068 was a 1984 delivery to Colorado Springs, Colorado.

Waukon, Iowa received this Ford L8000 chassis Seagrave, in 1982. The cost of a commercial truck in the class 8 or the heavy truck classification begins to approach the cost of a custom chassis. That's why you don't see many class 8 or greater commercial fire trucks. The cost of the custom features on a commercial chassis is usually only a few dollars, or within five percent of a full custom truck.

The delivery in 1969, to West Overland, Montana, introduced the new wider P style cab. The older style K cab was 76 inches wide at the back of the door and tapered to 72 inches at the door post. The new P cab was a flat side cab, a full 80 inches wide. Notice the jog in depth from the side step to the pump panel face. Overall the cab also dropped 3 inches in height to 96 1/2 inches.

The P cab also brought a change in the pump. The new P pump used ball bearings to support the pump shaft instead of brass bushings. The change meant less resistance allowing more output from the same size engine. The partially completed chassis, at left, highlights the P cab configuration. Notice how the upper frame channel is flared out. The accessories of this power plant utilized all the space provided by the Z style frame and the hump in the engine cover. The configuration of the engine cover provided the widest possible engine compartment and an adequate width seat.

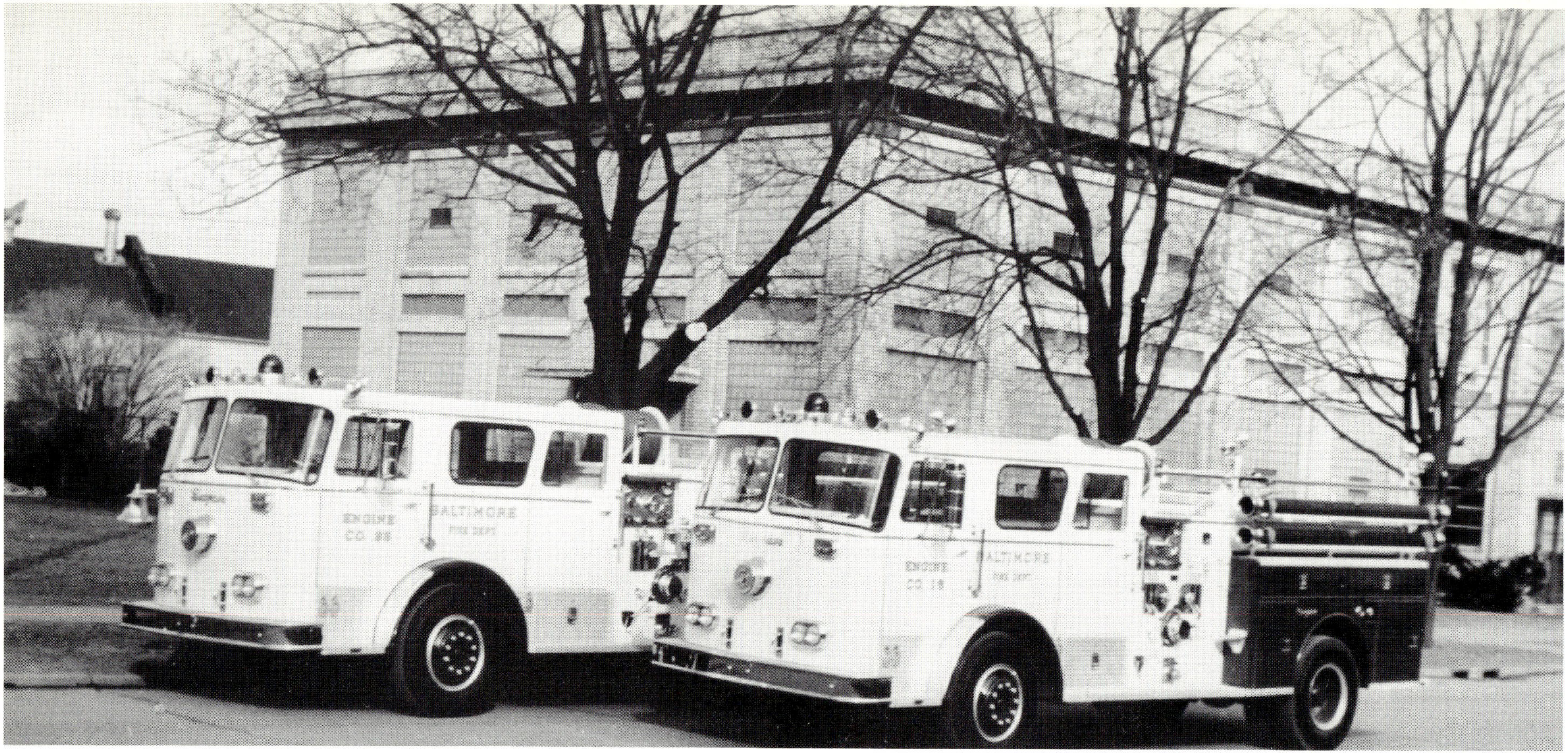

The 1971 deliveries to Baltimore, Maryland, were modified P seven man cabs. The opened rear doors show facing jump seats. Seagrave, long a proponent of enclosed apparatus, constantly modified its standard cab to provide maximum protection for the greater numbers of crew.

The back drop for this photo, above, is the administrative office building at Seagrave. The first floor corner office directly behind the apparatus is the president's office. Sales and accounting staffs occupy the rest of the first floor. Engineering for pumps and ladders is on the second floor of this building. Records and service departments are in another building.

The P cab design, like this 1969 delivery to Latham, New York, was Seagrave's base cab from 1969 to 1982. That eleven year span saw many changes to the driveline components at Seagrave. The 1976 pumper to Latham was a Bi-Centennial Special. It was painted red on the bottom and blue on the top with white stars. The following year the top was repainted all white.

This cab was designed to fit around a gas or diesel engine, but the gas engine was rapidly loosing favor to the diesel power plants. The venerable V-12 was used in only a handful of these cabs. The diesel engine provided more usable horsepower per dollar spent on fuel. It was out of service for tune ups less, and it was cooler running than a similar sized gasoline engine. To run efficiently, a gasoline engine must burn 14.7 pounds of air to a pound of fuel, either at idle or full throttle. A diesel engine on the other hand may run as lean as 80 pounds of air to a pound of fuel at idle, a considerable savings.

The reduced heat output was welcome by those who had to ride in the jump seats. A side benefit to the diesel was reduced radio interference. The dual ignition set up on a V-12 engine could cause a lot of radio interference.

The P style pump, an updated version of the old K style Seagrave pump, (see page 283) was replaced by a fire pump from the Waterous Company, St. Paul, Minnesota. The old style Seagrave pump was limited in its capacity and its castings were produced by a local foundry, not Seagrave. To increase pump volume, past the 1,500 gallon range, a new pump would have to be designed or purchased. In order to take pump quality and pump volume beyond the standards of the day, Seagrave worked with Waterous to produce the new family of pumping engines. Waterous was already a supplier to the FWD pumpers provided at the Clintonville facility.

The S cab, a lower version of the P cab, was produced from 1971 to 1982. The S cab was designed to lower the overall height required for rear mount aerial deliveries, but it soon found its way on a few pumpers as well. Serial #R4535 went to Lindenhurst, New York.

This H cab and the W cab were produced from 1977 through 1990. These cabs were characterized by the large amount of glass they contained. The jog in the window line, for the vent window, is the easiest way to spot the cab series. While the new series cabs may look similar, no major cab components are interchangeable with the previous model cabs.

The first W or low cab was shown at a convention at Denver, Colorado, in 1977. The show special was repainted and delivered to Detroit, Michigan, in December, 1977, at a cost of $83,000. The taller version or H cab, like the deliveries to West Deer Township, Cheswick, Pennsylvania, and New Orleans, Louisiana, started coming off the assembly floor in mass, in 1978. There was about a 10 inch difference in overall heights between the H and W cabs.

Parallel with the introduction of this cab series was the dominate change-over to automatic transmissions in the driveline. The automatic transmission was more forgiving of driver abuse than was a clutch and standard gear box. In cities, the automatic transmission allowed more driver concentration on traffic congestion. In volunteer departments, where drivers may very well be inexperienced, the automatic transmission provided an extra measure of convenience and security.

If a community desires a four wheel drive fire apparatus, what better place to come than the FWD Seagrave. The delivery to Prentice, Wisconsin, carries both the FWD and Seagrave name plates, because it has the four wheel drive option. The Clintonville factory produces the powered steering axles for four wheel drive apparatus.

The Customer Delivery Center is shown here. Many customers prefer to come to the plant to inspect their apparatus upon completion. Any customer concerns are more easily addressed at this time than after delivery. Customers may arrange for delivery or drive the apparatus home. The vehicle shown was sold to Anacortes, Washington. It is awaiting final inspection at the Customer Delivery Center.

The J cab, or flat front cab, was first introduced in 1988. It's low profile twin, the L cab debuted in 1990. These flat front cab forwards harken back to the first motorized Seagrave deliveries. They too were flat front, driver ahead of motor, fire apparatus.

The Seagrave J cab is distinguishable from other Seagrave cabs by the vertical air intake just behind the front door. The L or short J cab is used on ladder trucks only (see page 338). The construction of these units is the same except nine inches is cut from the J cab belt line to produce the L cab.

The top two photos show typical J cab pumpers. Dearborn, Michigan, received their apparatus with a top mounted control panel. Dearborn, home to Ford Motor Company is making the switch to custom fire apparatus.

The other top photo is a 1990 delivery to Windsor Ontario. Windsor, like so many communities across North America, had Seagrave apparatus years before, then shopped competition, only to return to Seagrave. This pumper protects the same street that was shown in the photo on page 71.

The Clintonville Fire Station is the backdrop for a photo of the new J cab apparatus. The J cab is not just a flat front H cab. There are major differences from the outer skin to the mechanicals. For example, the doors on the J cab are full length versus the short door on the previous H and W series cabs. Also, the radiator on the H cab is mounted crosswise on top of the frame rails. In the J cab, the radiator is mounted inside the frame rails for a slightly narrower and lower profile.

Current apparatus from Clintonville, sport less trim and stripe than previous deliveries. There are many reasons for this. One is today's fashions in cars, as well as trucks, call for less trim. Picture today's Lincoln against a 1958 Buick, for example. Another reason is over the road delivery. Many of today's apparatus are delivered with the stone shield, shown right, just to protect the paint.

Seagrave has a decorator ready to stripe an apparatus, but stripe and lettering damaged in shipment could be hard to match locally. So the plant prefers customers have trim done locally.

J 7 8 0 5 6

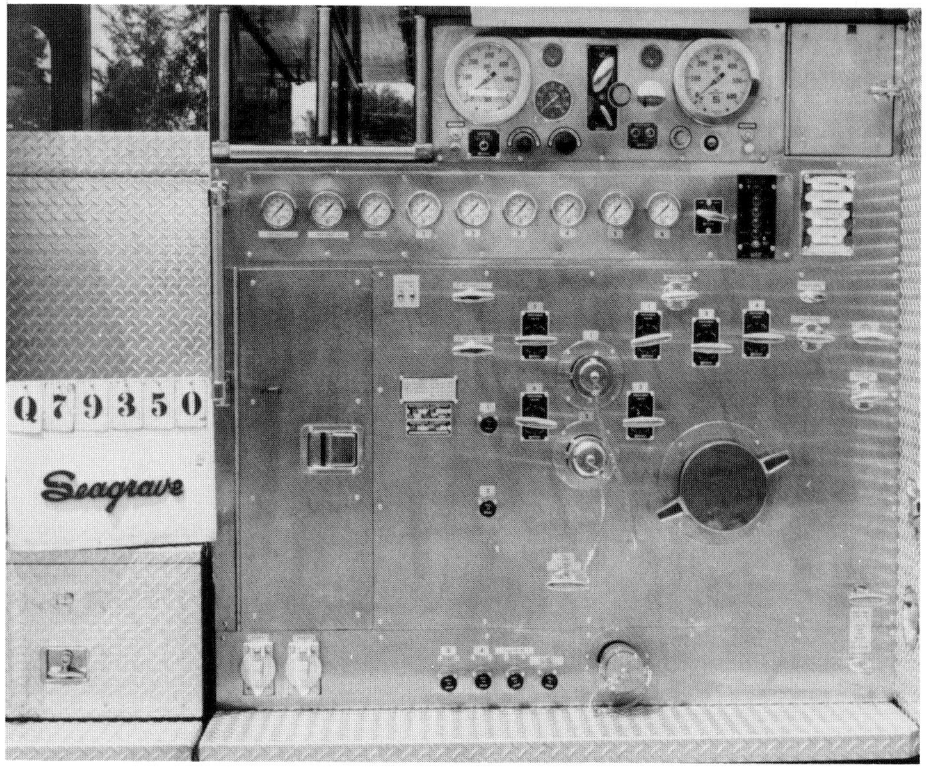
Q 7 9 3 5 0
Seagrave

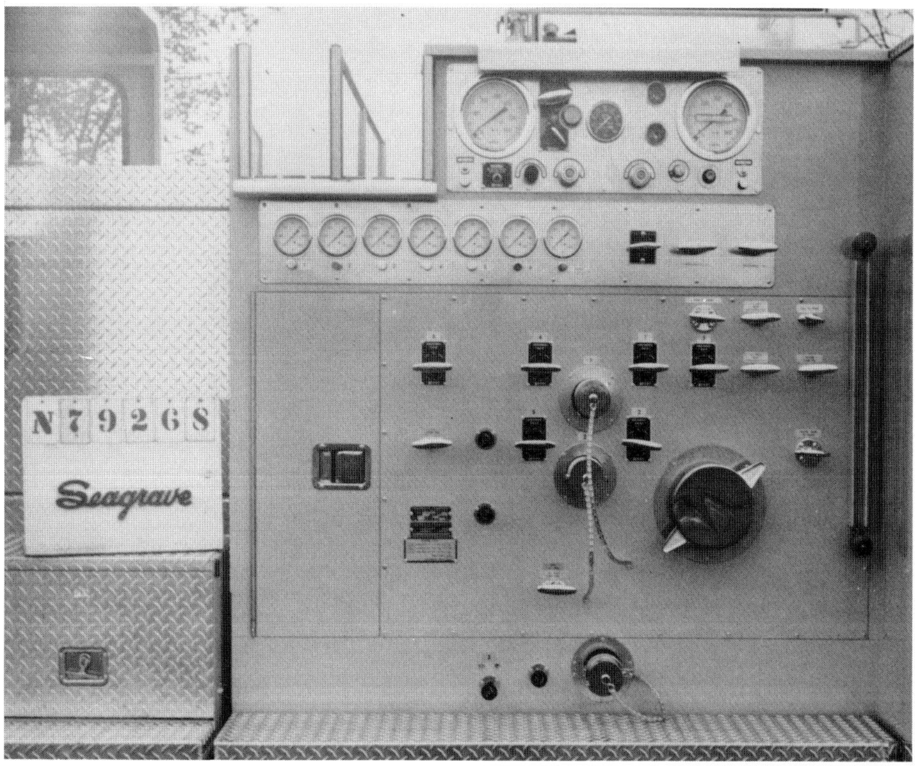
N 7 9 2 6 8
Seagrave

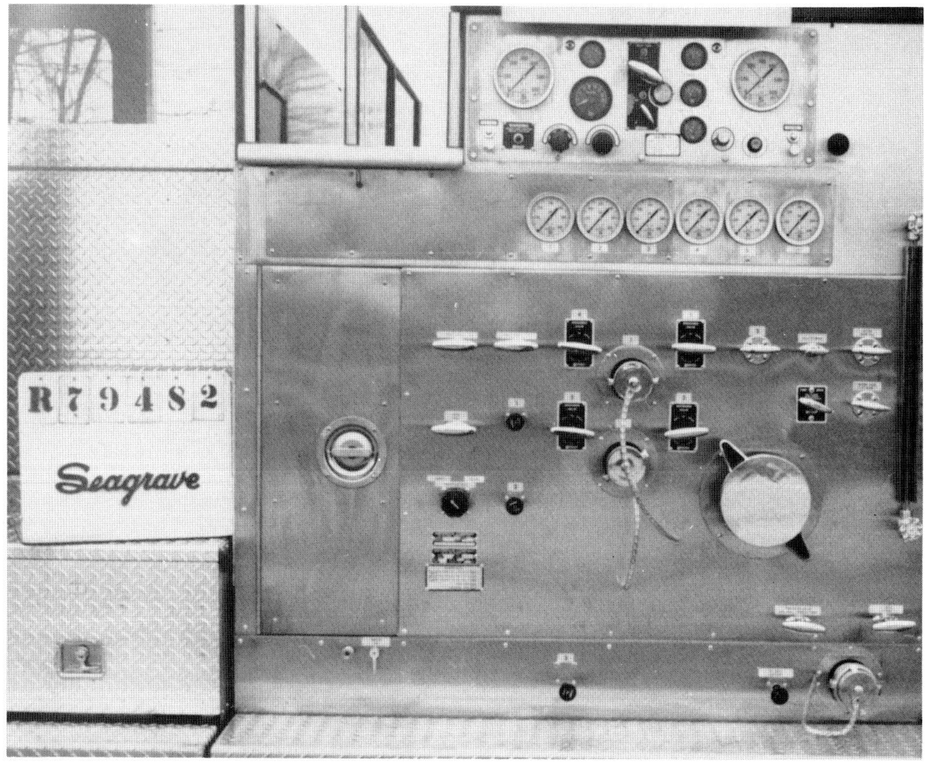
R 7 9 4 8 2
Seagrave

Pumpers delivered with the old K and P model pump had discharges located over the pump intake, as shown on page 314. The newer Seagrave deliveries with the Waterous pumps have dischargers mounted vertically next to the pump intake. The diversity in pump configurations is typified by the photos at left:

1979	J	78056	Corvallis, Oregon
1982	N	79268	Santa Rosa, California
1983	Q	79350	New York Mills, New York
1985	R	79482	Colma, California

Pumping engines are typically categorized as long or short wheel bases. The length of apparatus many times is dictated by the station house to which it will be assigned. The utilitarian aspects of hose load or water tank size is usually the other prime consideration. Wheel bases can be as short as 172 inches for a plain five man cab unit to over 200 inches for a special. The J cab deliveries shown are for Brainerd, Minnesota, above, and Columbia, South Carolina, right.

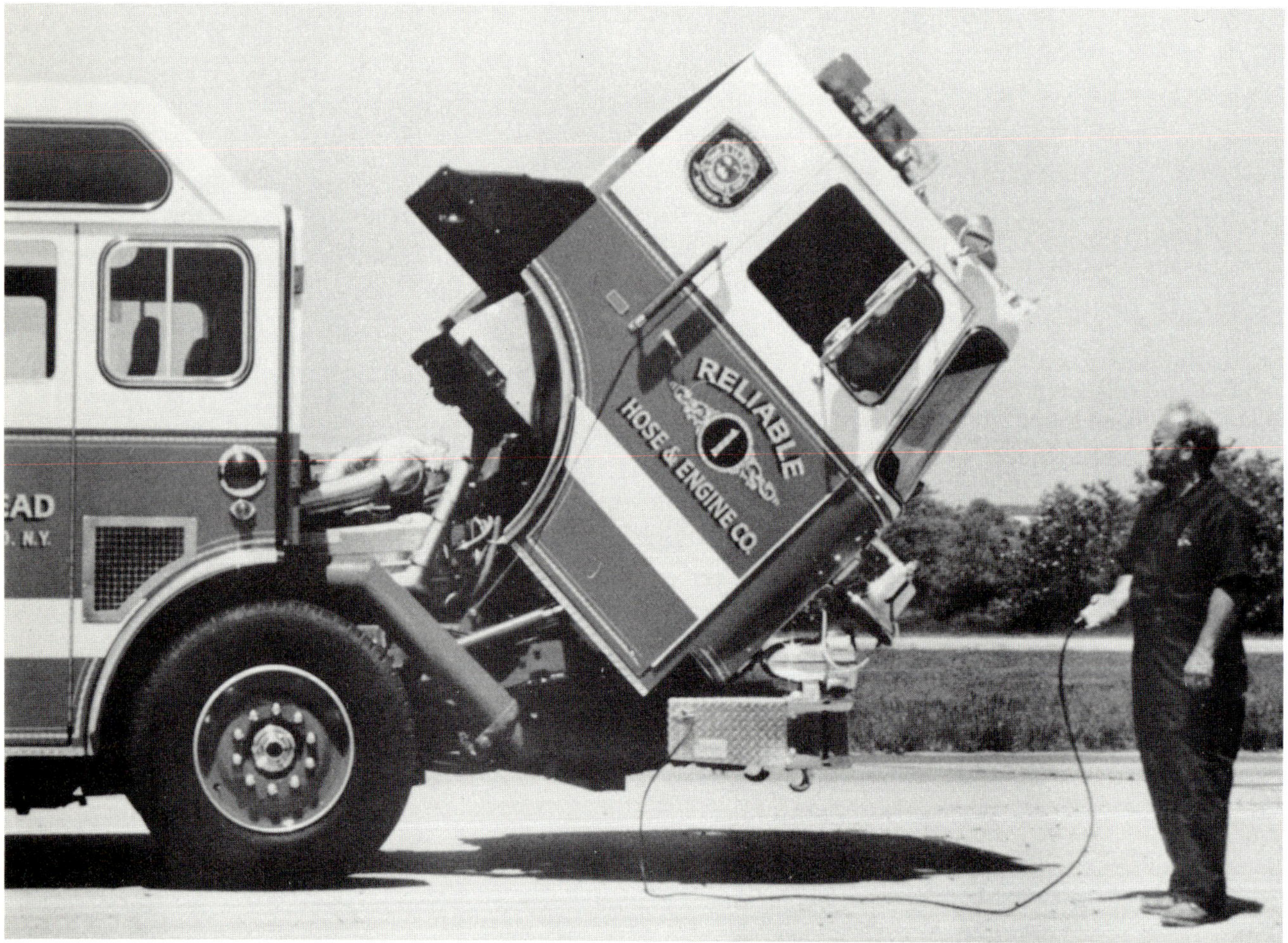

The look of tomorrow from Seagrave is the new custom tilt cab. This T cab design became available in 1991, in six or ten man seating arrangements. The six man cab is a full tilting cab, the ten man cab is split and the only the front section tilts. Either cab gives easy access to the major mechanical components. Tilting the cab is by an electric over hydraulic system. Checking of coolant and fluids is made through access openings in the drivers compartment.

Many departments order these large cabs to provide a safe haven for the firefighters during times of adverse weather or social climates. The ten man version may be adapted to become a six man cab, in conjunction with a first aid station or a command post.

This cab configuration is used in conjunction with an engine forward apparatus. The engine is actually located between the driver and officer. The engine forward feature allows for a more open crew area. The open walk thru crew area can be utilized for many purposes. Having all firefighters in the same enclosure means the radio communications can now be heard by the entire crew.

All custom pumpers are designed and built by Seagrave, providing the customer a single source responsibility. Fiberglass water tanks are standard on pumpers, but other options are available. The pump area is free standing from the hose box. This last feature eliminates stress cracks as the hose area is subject to different loading stress from the rest of the apparatus.

This new T cab and the four door J and L cabs will be the designs well into the next century. The two door, open jump seat, cab is obsolete. As a historical fact, this engine forward, flat front cab is exactly the same style as the first Seagrave motor delivery, in 1907.

Seating Arrangements

6-Man

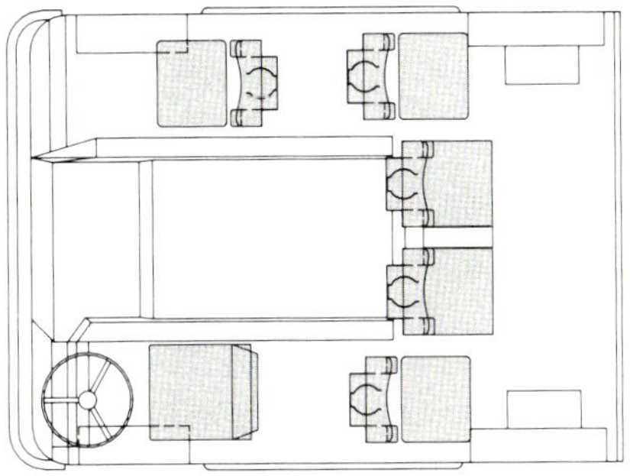

10-Man

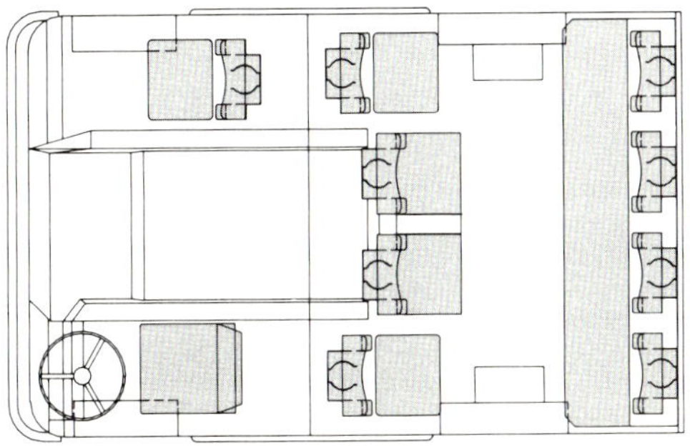

With Optional Medical Compartment

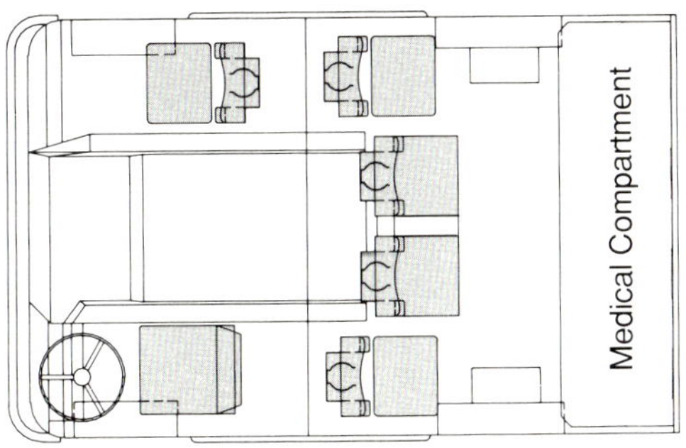

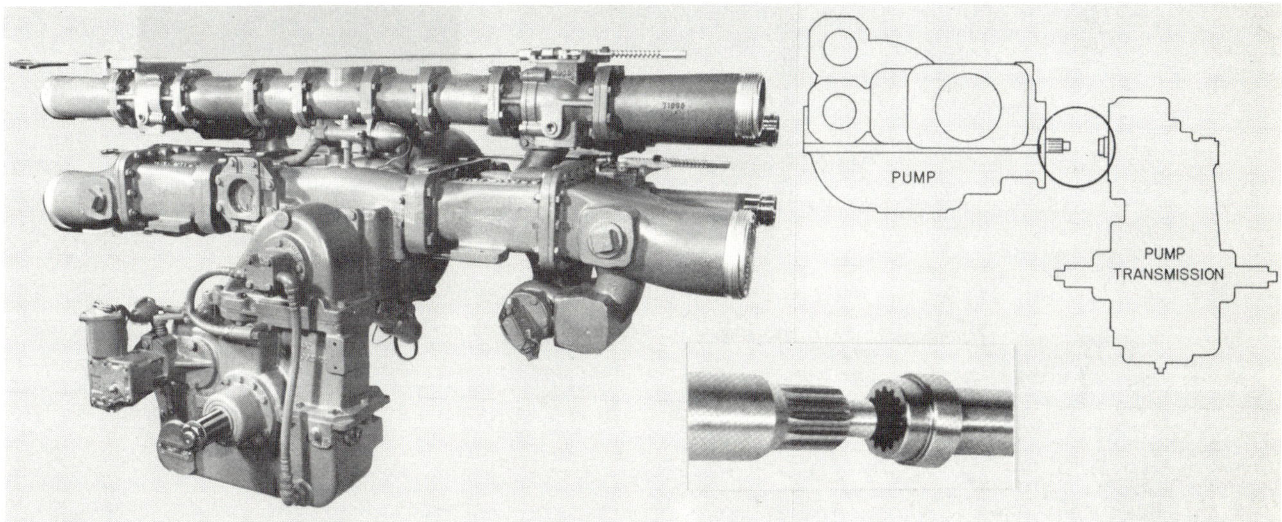

In 1973, Seagrave began to use pumps from the Waterous Company, South St. Paul, Minnesota. The old style Seagrave pump castings had been produced by a local foundry since the move to Clintonville. The off-site foundry location caused quality concerns. The old pumps needed updating to be able to consistently produce 1,500 gpm or better. Seagrave chose to offer the Waterous pump line rather than reengineer the older Seagrave design.

Waterous built their first pump, in 1886, for a steam fire engine. Their first gasoline driven pump delivered 175 gpm, in 1898. In 1907, Waterous delivered a gasoline pumping engine to Alameda, California, that used the same engine for road or pump power.

Waterous pumps are noted for their performance, but there are three other features that make Waterous a desirable pump. One - the pump case is split so the bottom half of the pump can be removed in the vehicle. Two - the pump can be removed separate from the pump transmission. Three - the pump transmission employs an internal chain to transmit power. The chain is efficient and quiet.

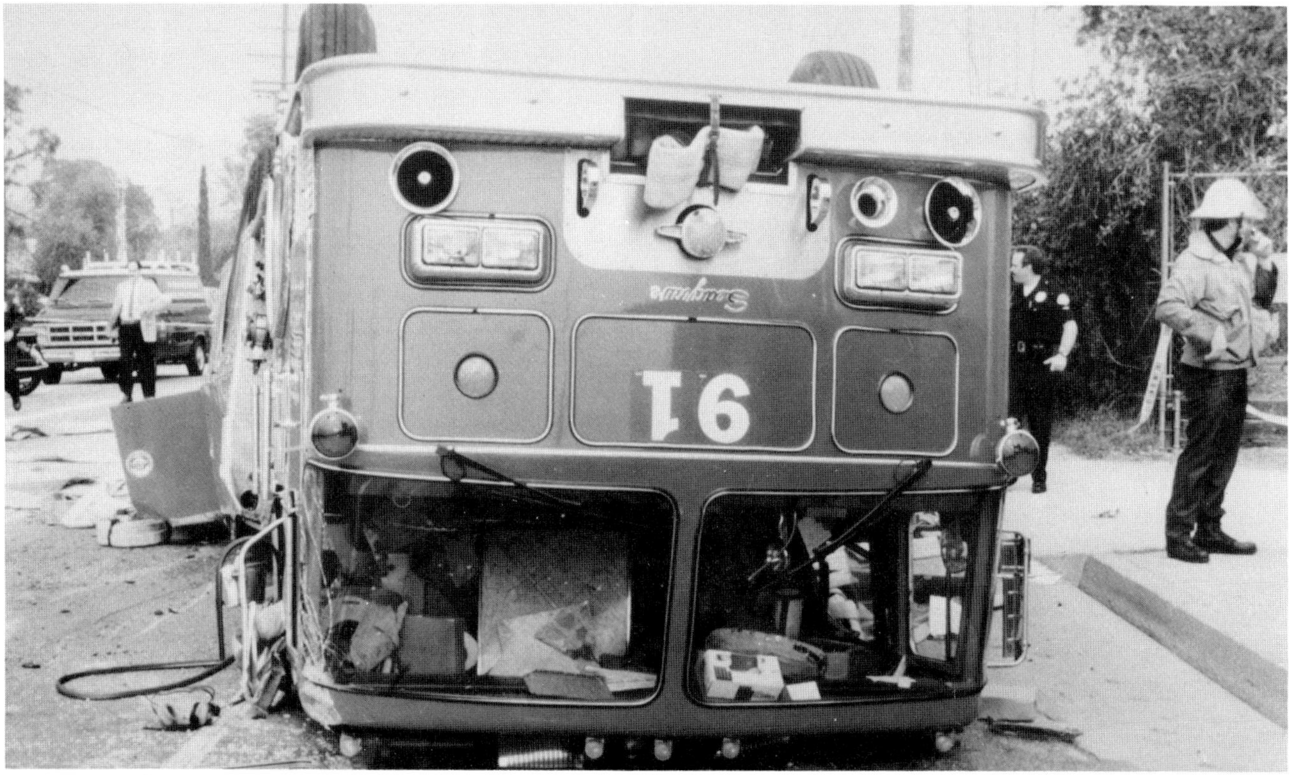

SERVICE FEATURES

ROLL OUT BATTERY TRAYS

Eliminates the hard labor of checking servicing, removal or installation of batteries. Heavy duty rollers for effortless operation. Sturdy locking latch and handle for positive security when traveling.

HINGED SWITCH PANEL

Large sized rocker type switches with integral pilot lights. Non-reflective background assures positive identification. Swing-down feature simplifies servicing of the switch system.

FRONT ACCESS PANELS

Provides easy access for servicing of wiring, pneumatic brake lines, heater and defroster vents and windshield wiper system. Panels are easily removed and replaced by use of "zip" gasket.

Sometimes it's the things you can't see that make the difference, like Seagrave's built-in roll bar protection system. Proven in 1936, and many times since: stay in the vehicle, and stay protected by its structural integrity.

Los Angeles' Engine 91 was broadsided by a 16 year old, allegedly driving a stolen vehicle. The five firemen in the cab walked away. One firemen received several stitches, the others were returned to duty. A firemen was quoted as saying that the scary thing was "there I was sitting upside-down knowing that when I released my seat belt, I'd fall on my head."

In Gary, Indiana, a new J cab pumper hit and sheared off a utility pole. The skeleton of the cab showed though the skin, but the serviceability of the vehicle remained intact. The pumper remained in service for two months after the accident before being sent out for repairs.

Some of the features that you can see are shown at left. Pop out, roll out, and lift out access to vital components. Close examination of the wire in the harness would show the path of the wire printed right on the wire. Example: gas tank to gauge, or #128 siren.

It's these hidden extras that keep Seagrave custom rigs in service, often rough urban service, long after other makes of apparatus have been placed in reserve.

Seagrave fire equipment is found in large cities the world over. It is designed to exceed the rigors of urban use. Smaller communities benefit from these rigid design requirements. An apparatus that can operate 20 or more years in a large city should be more than adequate for most communities.

The Clintonville factory has a railroad siding. The federal government is prone to ship by rail. But, the vast majority of goods, in and out of the plant, are driven or trailered. Real truck fans should enjoy these two photos, taken at the plant gate 27 years apart. A quad, in 1964, going to Euclid, Ohio. And, a rebuild coming in from Patillian, Nebraska.

Rail shipments lost favor over the years because rigs broke loose in the rail cars and rolled back and forth for hundreds of miles. The classic story is when Fort Worth, and El Paso, Texas, each ordered rigs at about the same time. El Paso received their rig a year before Fort Worth. The El Paso delivery was over the road. The Fort Worth truck made the trip twice by rail, then once over the road.

The midship mounted aerial, like its predecessor, the service ladder truck, were in less demand in later years. The addition of a pump made the configuration a little more saleable, but the sheer overall size of these Quads limited their use. Small communities with fewer traffic jams, championed the versatility of the midship aerial quad.

Shown here are; serial #P3560 - a 1963 delivery to Manhattan, Kansas and serial #N180 - a 1961 delivery to Mundelein, Illinois. With the passing of the P cab style in 1982, no more open cab apparatus were produced.

By 1963, the price for a big V-12 powered, cab forward, 100 foot aerial was $43,800. A 1,000 gallon pump and equipment was a $6,000 option. The enclosed cab was a $650 option. The K style cab was only $500 more than a similar anniversary series.

Glenside, Pennsylvania's, model K, 100 foot aerial is shown attacking a fire at a local eatery, in 1967.

Chattanooga, Tennessee's 1982 pump and service ladder combination was assigned to a company that had responsibility for many two story apartment complexes. It had a 1,250 gpm pump, 500 gallons of water and a complement of extension ladders.

To defend its missile sights, the government of Saudi Arabia purchased seventeen of these sand colored Quads. They were four wheel drive equipped units that featured 75 foot aerials, a small booster pump and dry chemical extinguishers.

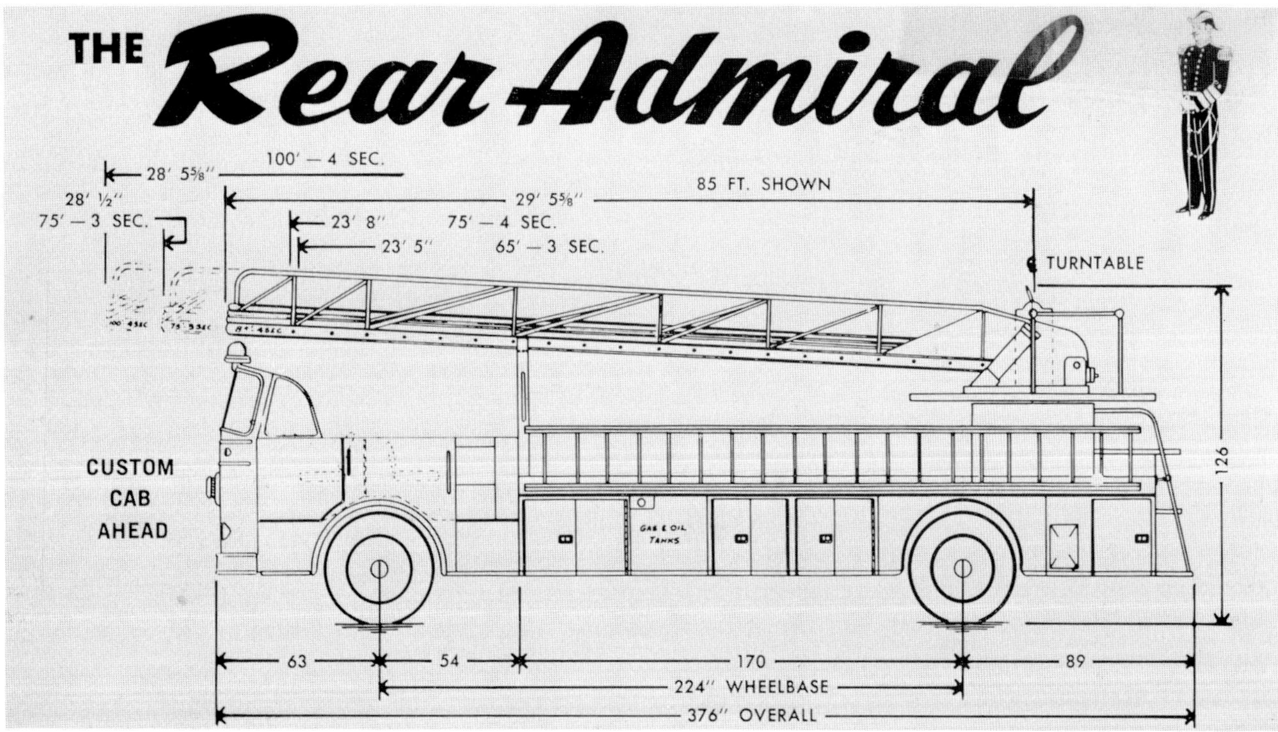

For years, the term Rear Admiral was synonymous with a rear mounted aerial ladder. It almost became a generic term like Kleenex or Hershey Bar. The Rear Admiral, introduced in 1965, was designed to carry the longest aerial ladder on the shortest possible wheel base.

Overall length was the big sales feature. A Rear Admiral was almost ten feet shorter in overall length than the midship mounted aerial of equal length.

The early Rear Admiral ladders also featured rear wheel steering. The rear steer was to be used for spotting the ladder at the fire scene. The driver was to engage the rear steer at 5 mph or less. The rear steer option was not cost efficient and it was not offered after 1967.

The early rear steer trucks were easy to spot by the large single rear tires mounted on wheels and hubs similar to the front axle. The rear steer, Rear Admiral delivery to Chicago's O'hare field was serial #72592, a 1967 delivery.

The basic layout of the rear mounted ladder/tower was not new to Seagrave. Its water towers followed this configuration years before. It should be noted that the water tower deliveries had hard rubber tires and wider than normal rear axles to provide stability. For the new Rear Admiral, Seagrave opted for "A frame" stabilizing jacks. These stabilizers extended 150 inches, center to center.

The first Rear Admiral delivery went to Beverly Hills, California, in 1965. The first of the new style rear mount ladder, a 1987 delivery to Northlake, Illinois, is shown above, for comparison. The old style ladder would support a 200 pound load on the tip, when fully extended horizontally, but could not lift it. The new style ladder will lift a 250 pound load.

The Detroit Diesel engine and Allison automatic transmission were becoming a favorite driveline. The automatic transmission was standard equipment on the early rear steer, Rear Admiral because steering that rear axle was done via foot pedals.

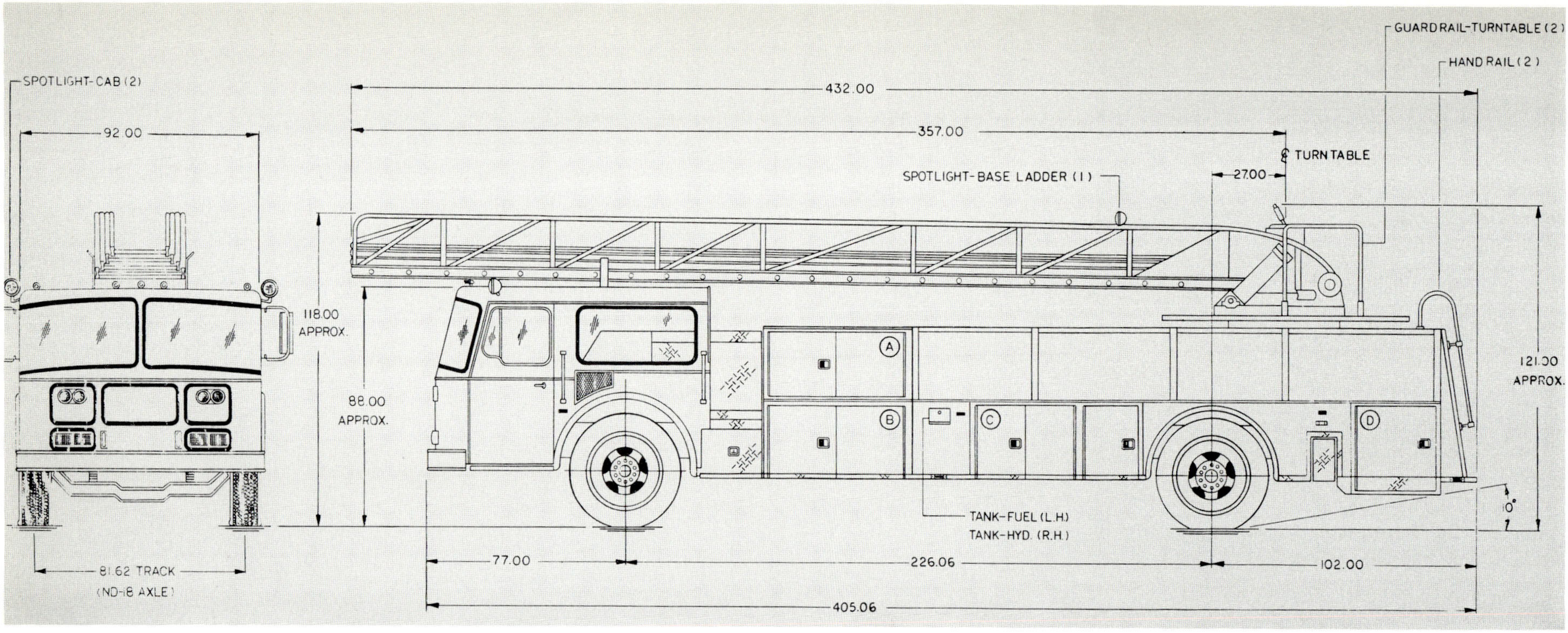

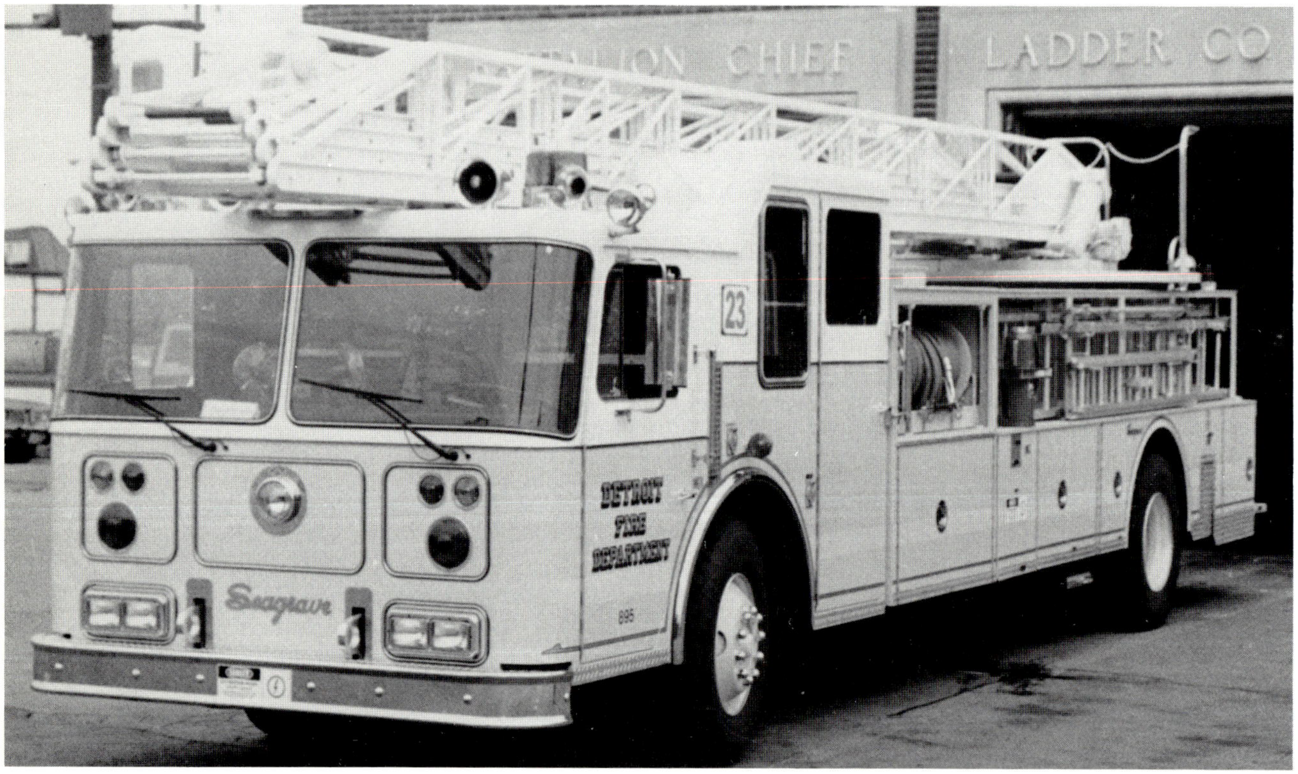

The drawing, above, gives the basic layout for a long wheel base, rear mount aerial. The overall length and height of fire equipment keeps growing, but the station houses often limit the size of new apparatus. The tallest point on the apparatus, above, is the operation levers which just tops ten feet. Another critical dimension is the ramp angle of the apron. The 10° exit angle allows liberal clearance for the tail board as the truck leaves the station.

The 1990 Detroit, Michigan, ladder cost $286,276. That's a tad more than the first Seagrave ladder truck assigned to Ladder 23 (see page 194). This brings the total Seagrave delivers to Detroit to 212. Many cities, large and small, have been repeat customers: Chicago, Illinois, has received over 150 units; Los Angeles, California, 200 units; and New York City, over 400 Seagrave apparatus.

A ladder truck's main function is a life saving device. The engine may put out the fire, but the ladder truck saves lives. So often we look at an aerial ladder truck and fail to recognize that this truck also carries other equipment: tools, life nets, breathing apparatus, and most importantly, ground ladders. The 1981 delivery to New York City, even came equipped with pompier ladders. New York rear mount aerial ladder trucks, through 1980, were identifiable by the extra cab compartment on the side. New York used a crew of an officer and six men on one ladder truck. The low seniority man often rode alone in the side cab.

The rear mount, at left, went to Davenport, Iowa, in 1983. The Davenport truck has a W cab. The New York truck has a S cab. The difference in cab width is evident when comparing these two photos.

The Cab

The RA110, Rear Admiral, 110 foot aerial, ladder truck was delivered with the H/W style cab, until 1990. Now it is available with J/L series cabs. The earlier cabs had a more rounded front and just slightly more (about 4%) windshield glass. Seagrave cabs still have the famous roll-over protection bar. Cabs are available in galvaneal steel or stainless steel. Seagrave uses aluminum on the floor of some cabs as long as its use does not effect the structural integrity of the cab.

Shown on this page are the W and L cabs for comparison. The trucks on the opposite page have the J style high cab. Both the J and L style cabs measure 94 inches wide. Their flatter front gave the engineers an extra three inches of equipment space. Cab interiors are finished in black. The exterior is painted with Dupont Imron® or P.P.G. Durethane®.

The Ladder

The latest innovation in ladder technology is the RA110 aerial ladder from Seagrave. Rectangular tubing is used in the construction of the main and truss ladder beams. It has a working length of 110 feet from the ground. The mechanism of the ladder weighs 7,000 pounds and has a working load of 550 pounds.

The Seagrave ladder on the previous pages (336-337) were classified as 200 pound tip load ladders. This ladder is rated at 550 pound tip load. The four section ladder is trussed three ways, vertically, diagonally, and horizontally. To stabilize the ladder, "out and down" style, outriggers are provided. The total distance between the outriggers is 16 feet.

The increased stability of this new ladder allows for the use of an optional 4 inch prepiped waterway.

The photo at left shows East Moriches, New York's 75 foot Rear Admiral during a demonstration, in 1985. This Rear Admiral version on a single rear axle called for separate jacking procedures (notice the hydraulic jack under the front end). Later versions mounted on tandem rear axle chassis did not require the extra front axle stabilizing. The outrigger rear axle stabilizers have an overall spread of 17 feet 4 inches.

The base of the aerial ladder is free of the cable drum of previous aerial ladders. The extending and retracting of the aerial mid and fly sections are now controlled by twin hydraulic cylinders and cables. The base section of the three section ladder is 31 inches wide. The mid section is 27 inches wide and the fly section 22 inches wide. That's two inches wider than the older style ladder. The hand rail portion of the truss ladder is supported at every other rung. Optional with the new Rear Admiral is a built-in four inch waterway.

The Hales Corners, Wisconsin, 1990 delivery features a tandem axle arrangement. Typically, people would think that big cities would order big trucks. In many cases the potential user for a truck like this is a smaller community. An apparatus like this with an aerial ladder, a 1,000 gpm pump, a 1,000 gallon water tank and a complete set of tools is really a fire station on wheels.

If one volunteer drives the apparatus to the fire scene and the other volunteers meet at the scene they have their station with them. There's even a roomy cab in which to suit up.

The Squrt® style boom on a Seagrave chassis is a viable option for a community that wants an elevated master stream at a thrifty price. Squrt® is a trade name, an articulated boom made by the Snorkel Company.

Shown above is a Tele-Squrt® on a 1980 Seagrave P cab chassis for the Cincinnati, Ohio, Fire Department. To the right is a high J cab for Adelanto, California, and an L cab for Dearborn, Michigan.

The low L cab allows for an overall height of 123 inches. The Dearborn engine also features stainless steel open lattice (rough edge facing up) running boards. It's going to be tougher to sit on the tail board and swap stories from now on.

The use of Tele-Squrt® equipment allows a relatively short pumper to also act as an elevated water way and an emergency ladder.

The 50 foot combination water tower-ladder, with its outriggers, is added, as a unit, to the end of existing frame rails. The assembly requires a hose bed space of 18 x 28 inches, and weighs 5,100 pounds.

Young Spring and Wire Company of upper Sandusky, Ohio built these three piece booms for Seagrave, and installed them on Seagrave chassis. Kentland, Maryland received this first delivery of the Seagrave Eagle, in 1963. The boom was all steel, but the platform was aluminum. A four inch waterway was built into the boom. It could reach heights of 90 feet and required about three minutes to raise the boom and rotate it 360 degrees. The apparatus was 40 ft. 5 in. long. A competitor's two section boom unit was more than 8 feet longer.

Seagrave had delivered aerial ladder platforms since 1929 (see page 199). The two boom Eagle model delivery to Nashville, Tennessee, a 65 foot model, (see page 292) and the 90 foot eagle model, shown here, were hard to maneuver in traffic because they were top heavy. The three section boom weighed 11,000 pounds by itself. The concept needed more refinement.

Further developments included the solid boom style produced by the Trump Company of Oliver, British Columbia, Canada, and the Grove telescoping boom shown above. The Trump platform truck went to Greenville, South Carolina, in 1972. The Grove platform went to Eauclare, Wisconsin, in 1974. Test research on aerial platforms remained ongoing for years. The result of decades of development is highlighted on the next page.

The latest in a long line of innovations from Seagrave is a 105 foot aerial platform quint, the Apollo 105. The all steel boom and all steel platform exceed the current standards for aerial platform apparatus.

The pump is a Waterous 1,500 gpm, two stage, pump capable of discharging its rated capacity at 150 psi or one half its capacity at 250 psi. The standard water tank is a 150 gallon capacity and is attached to the pump with a 3 inch pipe.

The aerial platform waterway is prepiped with a telescoping aluminum tube. The base section is five inches in diameter. The middle section of 4 1/2 inches slides into the base. The fly section is a 4 inch diameter tube. Water at the platform can be discharged through the turret gun and by a shower nozzle mounted under the platform.

The 105 foot aerial platform should revolutionize the way many communities fight fire well into the next century. With this large capacity combination you can lay 1,000 feet of five inch suction hose from the hydrant to the fire scene. The large volume water system will provide a 1,000 gpm elevated stream and enough reserve capacity for deck guns and hand lines.

The 105 aerial platform provides the safety of a full ladder and the convenience of an aerial platform. In a multiple rescue situation some victims could be escaping down the ladder, as others are stepping onto the platform. An ordinary aerial platform, without a ladder, would have to make multiple elevations to the rescue areas. This fire station on wheels takes less space in the station than most tractor-drawn aerial trucks.

Major cities on each end of the country had different specifications for their aerial ladders. The upper photo is a 1961 order for Los Angeles, California. The other delivery is for New York City, in 1962.

The Los Angeles delivery still featured wood ground ladders, and the three section aerial ladder. New York specifications called for a four section ladder with tapered, rather than the standard round end, hand rails on the ladder. Departments typically ordered either no hand spot lights, or a set of spot lights. New York specifications call for a spot light on the officer's side only.

The dash of the L. A. rig has a button and gage for just about everything. The long handle under the right side of the steering wheel is for independent operation of the trailer brakes.

The upper right photo shows the longer than standard aerial ladder control levers, a Los Angeles requirement.

The other two photos show a New York aerial ladder test at the Columbus Fire Department Shops, May 23, 1962. The pilot build for New York City's order is subject to a review before the total order is completed. New York representatives witnessed this test, and reviewed adherence to F.D.N.Y. specifications.

Part of the test included using two deluge guns at high pressure. Twin pumping engines scheduled for delivery to Wichita, Kansas, provided the water. Other tests included maneuvering the ladder under load. Two steel weights, totalling 500 lbs, were attached to the base ladder during the timed tests.

At delivery, another set of checks is made regarding the air brakes, fluid levels, and ladder cable adjustments.

XCELSIOR H&L CO. No 1
ORGANIZED 1874
LIFE NET

LIFE NET
N2756
THE
Seagrave
CORP.

FREEPORT
FIRE DEPT.
4456
KEEP BACK 500 FEET

The view from the tiller seat is fantastic, up high looking over the traffic on a warm evening on the way back from a run. There is no better ride. Unfortunately that scenario happens only five percent of the time. When it's raining, cold or snowing, the comfort of an enclosed cab is welcome.

Shown for contrast, at left, are an open tiller on a 1962, 100 ft. ladder for Baltimore, and a 1961 delivery for Freeport, New York. The Freeport delivery had a roll back tiller compartment. This style, while protective, could also become uncomfortable because of the lack of controlled ventilation.

The improved designs at right, provided protection and optional ventilation. The solid roof meant the tillerman would always be seated and therefore less subject to mishap.

Shown right, are the basic options for the tiller cab. The photos also depict some of the ladder compartment enclosures available.

From one end of the country to the other, when it comes to maneuvering in traffic or on narrow streets, nothing is as versatile as a tractor-drawn aerial ladder truck. Over eighty years of making this design have culminated in the 1990 delivery to New Haven, Connecticut, and the 1991 delivery to Burbank, California.

The ladder looks somewhat like earlier designs, but there are subtle changes. The ladder rungs, for example, now have a corrugated rubber covering to prevent slipping. The turntable now has three ball bearing races. The usual two bearing races to take the ladder load is augmented by a third race to take the ladder reaction load. The complete ladder mechanism uses 472, 3/4 inch ball bearings.

The new 550 pound tip load ladder truck features four swing out stabilizer arms. The new arms are necessary to counteract the forces generated when the ladder has to lift a load from the horizontal position. The ladder has a safety factor built in and consideration is given to the possibility of an ice coated ladder.

The ladder is rated at 550 pounds tip load without a prepiped waterway and 400 pounds with a waterway. The new style Seagrave ladders, with the box beam construction, will actually lift heavy loads from the horizontal position, with the ladder fully extended. The older style Seagrave "I" beam ladder could only support (not lift) 200 pounds when extended horizontally.

The Montgomery County, Virginia, fire department received their new 110 foot aerial ladder, in 1991. This new model features the new wide tiller cab for more comfort and safety.

Ladder 20 of Boston, Massachusetts, is shown at work and Ladder 38 of New York City, is shown at play. Either way the big stick from Seagrave is a useful and impressive life saver.

The quality of the Seagrave is taken for granted by most departments. In times of need, it has always been there. Since 1936, the Seagrave all steel aerial ladder has been the standard of the industry.

The action photos are from: Baltimore, Maryland - Playboy Club fire - 1969; Detroit, Michigan - Buhl building fire - 1984.

The Detroit photo was taken just a few blocks from where the photo on page 11 was taken. Both photos, taken 100 years apart, show eleven men on a Seagrave trussed ladder.

ALLENHURST
F.D.

LUBBOCK
15
FIRE DEPT.

lphia Fire Academy
EDISON
TRUCK Nº1

ALLENHURST
ALLENHURST
F.D.

F.D.N.Y.
38
F.D.N.Y.

APPLETON
Seagrave
TRUCK CO.
1
F75115
Seagrave

Shown above, is prototype ladder testing and the new style 100 foot, 200 pound, tip load ladder. The Patriot style four section 100 foot aerial ladder is the replacement for the older style Seagrave ladder.

Shown left, are samples of the equipment that have earned Seagrave the title "Greatest name in fire apparatus." The variance in those deliveries show the standard of quality is a moving target.

What styles of apparatus will the fire departments in the next century have to choose from . . . ?

In pumping engines - there are two basic chassis.
One - a midship engine pumper.
Two - an engine forward pumper.

The J and L cab Seagrave pumpers are offering the traditional full width front cab and separate jump seat areas. The T cab, engine forward, apparatus offers a full crew area. Pump volume displacements will vary depending on the local needs and water supply of the community.

In ladder trucks - there are four offerings.
One - a three section, 75 foot, 500 pound, tip load ladder.
Two - a four section, 100 foot, 250 pound, tip load ladder.
Three - a four section, 110 foot, 550 pound, tip load ladder.
Four - a 105 foot aerial platform ladder with a 1,000 pound, tip load rating.

From this line up communities, large and small, will be able to pick a base configuration and modify it to suit the special needs of their local. Just as the communities, shown left, all choose a Seagrave rear mount aerial ladder, but no two trucks look the same.

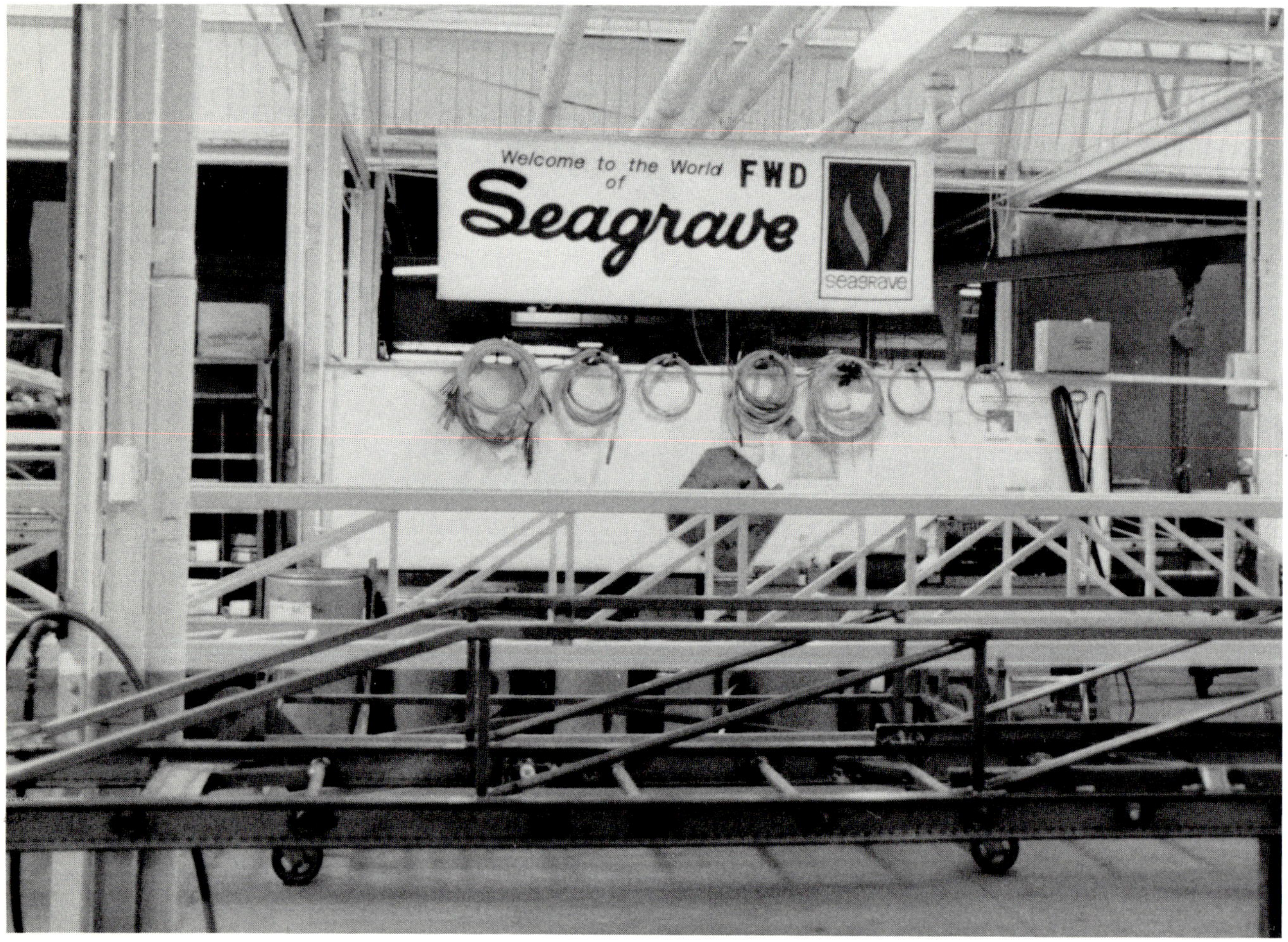

On a tour of the factory, the first room you enter would be the ladder assembly room, shown at right. The tour would take you through the machine shop, subassembly areas, the main assembly area, and out the back door to the testing area, shown above.

The ladders, shown left, are the two standard ladders in use the last few years. The standard ladder, a derivative of the 1936 style ladder. It has "I" beam construction, and the hand rail truss is supported every four rungs.

The 1978, and up, heavy ladder has boxed beam construction and the hand rail is braced every three rungs.

Although the standard ladder has never failed in use from defects, Seagrave is constantly upgrading the product.

The Seagrave plant does not have formal tours, but escorted tours are available. It's best to call ahead and plan appropriately for the tour. Dress comfortably and wear sensible shoes . . . it is a factory and a busy one.

The production of a new apparatus starts with the assembly of the basic frame rails, wiring, hydraulic lines and then suspension brackets. With the addition of axles, our build starts to look like a vehicle. The next stop for our rolling chassis will bring engine and pump to the assembly.

Sub-assemblies are done in separate buildings. After inspection, they are brought to the assembly area. This sub-assembly, off the frame, system allows for speedier delivery times. Cabs are available in galvaneal steel or stainless steel. The body and cab are treated with "Tectyl" in critical areas during assembly to prevent rust. Most purchasers of stainless steel cabs are concentrated along the sea coasts and Midwest states where salt is used to deice the roads.

These views show the partial apparatus mix in Seagrave's giant assembly facilities. There may be thirty or so pieces of apparatus in various stages of completion at any given time. Engineering and building apparatus for world wide markets mean extra quality must be built in, not bolted on, as it is difficult to make service calls in Hong Kong or Saudi Arabia.

Shown here is 90 years of the Seagrave aerial ladder. The first style spring raised wood aerial led to the standard steel aerial ladder, shown far left. A new style heavy ladder is shown, near left. The 105 foot aerial ladder platform is shown above.

There are nearly 850 photos and sketches in this book covering many communities. Seagrave however, had delivered apparatus to thousands of communities over the last 110 years and I could not show them all. The task of finding and documenting photos showing Seagrave's milestones meant excluding some apparatus.

Throughout this book, I have tried to let you explore each aspect of the photos. The text was only written to get your thought process started. There were many aspects of each photo to enjoy. The second, and successive readings of this book will uncover more points of interest.

I hope you enjoyed this touch of yesterday,
for a better tomorrow.

Matthew Lee